CLEARING OF INDUSTRIAL GAS EMISSIONS

Theory, Calculation, and Practice

AAP Research Notes on Chemical Engineering

CLEARING OF INDUSTRIAL GAS EMISSIONS

Theory, Calculation, and Practice

Usmanova Regina Ravilevna, PhD and
Gennady E. Zaikov, DSc

Apple Academic Press

TORONTO   NEW JERSEY

Apple Academic Press Inc. | Apple Academic Press Inc.
3333 Mistwell Crescent | 9 Spinnaker Way
Oakville, ON L6L 0A2 | Waretown, NJ 08758
Canada | USA

©2015 by Apple Academic Press, Inc.

First issued in paperback 2021

Exclusive worldwide distribution by CRC Press, a member of Taylor & Francis Group
No claim to original U.S. Government works

ISBN 13: 978-1-77463-350-2 (pbk)
ISBN 13: 978-1-77188-034-3 (hbk)

Library of Congress Control Number: 2014951596

Library and Archives Canada Cataloguing in Publication

Ravilevna, Usmanova Regina, author
Clearing of industrial gas emissions : theory, calculation, and practice / Usmanova Regina Ravilevna, PhD and Gennady E. Zaikov, DSc.

(AAP research notes on chemical engineering)
Includes bibliographical references and index.
ISBN 978-1-77188-034-3 (bound)
1. Gases--Cleaning. 2. Dust--Removal. 3. Vortex generators.
4. Scrubber (Chemical technology). 5. Separation (Technology).
I. Zaikov, G. E. (Gennadiĭ Efremovich), 1935- author II. Title.
III. Series: AAP research notes on chemical engineering

TP242 R39 2014   628.5'3   C2014-906571-X

Apple Academic Press also publishes its books in a variety of electronic formats. Some content that appears in print may not be available in electronic format. For information about Apple Academic Press products, visit our website at **www.appleacademicpress.com** and the CRC Press website at **www.crcpress.com**

ABOUT THE AUTHORS

Usmanova Regina Ravilevna, PhD

Usmanova Regina Ravilevna, PhD, is a professional mechanical engineer with over 15 years of experience in practicing mechanical engineering design and teaching. He is currently Associate Professor of the Chair of Strength of Materials at the Ufa State Technical University of Aviation in Ufa, Bashkortostan, Russia, where he supervises several PhD students. He is the author of over 200 articles and reports in the fields of fluid mechanics, fluid dynamics, separation, and working out of apparatuses for gas clearing. He is also a senior scientific employee of Academy of Sciences of the Republic of Bashkortostan in the laboratory, working on the project "The perspective designing and modeling of technological apparatuses." Dr. Ravilevana received his degree from the Ufa State Oil Technical University, Ufa, Bashkortostan, Russia.

Gennady E. Zaikov, DSc

Gennady E. Zaikov, DSc, is the Head of the Polymer Division at the N. M. Emanuel Institute of Biochemical Physics, Russian Academy of Sciences, Moscow, Russia, and Professor at Moscow State Academy of Fine Chemical Technology, Russia, as well as Professor at Kazan National Research Technological University, Russia.

He is also a prolific author, researcher, and lecturer. He has received several awards for his work, including the Russian Federation Scholarship for Outstanding Scientists. He has been a member of many professional organizations and is on the editorial boards of many international science journals.

He is the author of over 3,000 articles and reports in the fields of theory and practice of polymer aging and development of new stabilizers for polymers, organization of their industrial production, life-time predictions for use and storage, and the mechanisms of oxidation, hydrolysis, biodegradation, and decreasing of polymer flammability.

AAP RESEARCH NOTES ON CHEMICAL ENGINEERING

The AAP Research Notes on Chemical Engineering series will report on research development in different fields for academic institutes and industrial sectors interested in advanced research books. The main objective of the AAP Research Notes series is to report research progress in the rapidly growing field of chemical engineering.

Ali Pourhashemi, PhD
Professor, Department of Chemical and Biochemical Engineering,
Christian Brothers University, Memphis, Tennessee, USA

Ing. Hans-Joachim Radusch, PhD
Polymer Engineering Center of Engineering Sciences,
Martin-Luther-Universität of Halle-Wittenberg, Germany

BOOKS IN THE AAP RESEARCH NOTES ON CHEMICAL ENGINEERING SERIES

Quantum-Chemical Calculations of Unique Molecular Systems
(2-volume set)
Editors: Vladimir A. Babkin, DSc, Gennady E. Zaikov, DSc, and
A. K. Haghi, PhD

Chemical and Biochemical Engineering
Editor: Ali Pourhashemi, PhD

Reviewers and editorial board members: Gennady E. Zaikov, DSc, and
A. K. Haghi, PhD

Clearing of Industrial Gas Emissions: Theory, Calculation, and Practice
Usmanova Regina Ravilevna, PhD, and Gennady E. Zaikov, DSc

CONTENTS

LIST OF ABBREVIATIONS

MFD	method of finite difference
MFEs	method of final elements
MFVs	method of final volumes
RCs	rotary connections
RMS	mean-square error
VOF	volume of fluid
WAs	whirlwind apparatuses
WCDs	whirlwind contact devices

LIST OF SYMBOLS

ΔP_{dry}	hydraulic resistance of the dry device
ΔP_{g+l}	hydraulic resistance water gas a layer
A_r	Archimedes number
B	width of plates
C	an arithmetical multiplier/concentration of corpuscles
C_{cor}	factor of viscous resistance
d	diameter of the corpuscle/upstream end
D	diameter of the cylindrical channel/pipe
d_1	relative root diameter of an air swirler
d_d	diameter of a corpuscle
D_d	factor of their diffusion
d_{ekb}	equivalent diameter of cracks
F	area of free section of the screw channel
F_d and F_c	volume fractions
F_e	superposed force directly acting on a corpuscle
F_i	generalized forces
F_p	the relative area of alive section vortex generator, shares of a unit
F_r	resultant vector of forces acting on a corpuscle
h_l	height vortex generator
i	number of generalized coordinates
K	stream momentum
K_{ei}	factor of relative ekologo-economic hazard to pollutant
L_c	characteristic length
M	stream angular momentum
m_p	weight of the corpuscle
m_r	weight of a corpuscle
N	quantity of considered pollutants
q_1 and q_2	generalized coordinates and speeds of a corpuscle
r	radius of a vector
R	radius on an exit from an air swirler
r_3	radius of the central pipe
r_b	radius of a zone of quasifirm twirl

R_m	radius of the cylindrical chamber
t_d	time characterizing traffic of corpuscles
u	speed of a dust, equal to speed of gas
U_c	characteristic speed.
u_k	speed of substance k-й a secondary phase
u_m	ensemble averaged speed on weight
u_{mj}	projection of speed to an axis $x_j.m$
V_r	vector of absolute speed
W_φ and W_z	maximum values of the peripheral (tangential) and axial speeds of gas accordingly
W_2	relative speed of gas on an exit, mps.
W_{cp}	axial mean speed in the ring channel
W_{pl}	average speed of gas in cracks between plates
W_s and W_n	projections of speed of gas in the channel on coordinates s and n.
A	corner between the next plates on an input{entrance} in vortex generator
α_p	specific volume occupied with discrete corpuscles
γ	relation of density of a disperse and bearing phase
δ	thickness of the blade
E	minimal distance between plates on an output{exit} from vortex generator
ε_c	speed of a dissipation of a turbulent kinetic energy
ζ	factor of resistance of a drop
ζ_{dry}	factor of resistance of not irrigated apparatus
ζ_{ir}	coefficient of resistance, in terms of the changes which are brought in by an irrigation.
μ	dynamic viscosity of substance of a master phase
μ_c	viscosity of substance of a bearing phase
μ_m	ensemble averaged viscosity on weight
μ_c and μ	molecular and turbulent viscosities of a gas phase
ζ	coefficient of hydraulic resistance
ρ	density of a liquid
ρ_g	gas density
ρ_k	substance density k-й phases
ρ_m	mix density
ρ_n, ρ_l	density of gas and liquid
σ	a superficial tension

σ_ε	turbulent Prandtl number for the equation of a dissipation of a kinetic energy
σ_κ	turbulent Prandtl number for the kinetic energy equation
τ_{ij}	Cartesian components of tensor of voltage
υ	speed
$\acute{\upsilon}$	vector of speed.
υ_g	speed of a stream in the radial direction
υ_r	radial component of absolute speed of gas on an exit from a rotor
υ_z	speed of a stream along an axis
υ_φ	tangential speed of a stream
$\acute{\omega}$	angular speed of twirl
κ_c	a turbulent kinetic energy of a gas phase
Π	perimeter equal to the sum of the parties{sides} of a trapeze
p	mix density
P	pressure
T	corpuscle kinetic energy

PREFACE

One of the basic problems of the twenty-first century is environmental contamination by coproducts of chemical manufacturing. Protection of circumambient air against emissions of low-purity gas is an extremely topical issue. According to the United Nations, annually 2.5 million tons of dust is thrown out into the atmosphere. According to American ecologists, the quantity of the dust formed in the industry will annually increase by 4 percent at the expense of the general growth of industrial production. Under forecasts of the Ministry of Natural Resources of the Russian Federation, Russian-maintained reserves of oil will run out by 2015 and gas by 2025. By 2030, superiority will be occupied with coal and atomic engineering. Such changes in the fuel and energy balance in Russia and all over the world will demand perfection in gas-cleaning installations.

Reliability and overall performance of gas-cleaning plants of processes and apparatuses of applied chemistries depend on physical and chemical properties of corpuscles and thermodynamic parameters of heterogeneous medium. Concentration of firm corpuscles in gases and dispersion composition perfection of the organization of an operating procedure, which is an aspect of the technological apparatus and a regime of its work, depend on parameters of conduction of an operating procedure, features of the equipment, for example, from the way of crushing, drying, methods of rehash, and constructive characteristics of apparatuses. Many operating procedures are characterized by nonstationary regimes (variable in the time of concentration of a blending agent and charges of gases at change of the charge of a dispersion material).

One of the effective ways of clearing industrial gases from the weighed corpuscles is the wet mud process of the clearing, which has had of late years of considerable development in domestic industry and abroad.

Apparatuses of wet clearing of gases have the following advantages:

1. Wet dust collectors differ in rather less cost and higher efficiency of trapping of the weighed corpuscles in comparison with dry mechanical apparatuses.

2. Some types of wet dust collectors (turbulent scrubbers) can be applied to clearing of gases of corpuscles of a size to 0.1 µm.

3. Wet-type collectors not only can successfully compete with such highly effective dedusters as bag hoses but also can be used when bag hoses are not applied, for example, at a high temperature and a spray zone, at danger of ignitions, and explosions of cleared gases or trapped dust.

4. Apparatuses of wet clearing of gases simultaneously with the weighed corpuscles can trap vaporous and gaseous blending agents.

One of the reasons for restricted use of whirlwind centrifugal gas purifiers is the absence of reliable methods of calculation of gas kinetics and processes occurring in them, and also criteria of scale parameter from laboratory models to industrial installations. It results, therefore, that there are many working out of highly effective apparatuses, but their widespread occurrence is restrained by the absence of calculations and recommendations for realization of transition to necessary productivity and change of operating modes. Therefore, for designing new and effective utilization of known apparatuses for clearing of gas emissions, it is necessary to improve methods of calculation of gas kinetics and processes of separation of the two-phase twirled streams.

Defining characteristics of apparatuses for clearing of gas emissions are efficiency of separation and water resistance. Known methods of calculation of efficiency of the scrubbers based on the use of empirical functions, presenting parameters of fractional efficiency and dispersion composition of a dust, do not vary the significant accuracy. It is caused by the functions of mass particle size distribution of many industrial dusts which do not answer a logarithmically normal [lognormal] distribution because of the actions of several mechanisms of a dust formation. Use of some methods are inconvenient because of multistep, complexities, and labor contents of reception of initial data for calculation.

Despite these defects, wet gas-cleaning installations can be applied with success in the chemical, oil refining, and gas industries, and he black and nonferrous metallurgy, power engineering, and other industries.

Wet-type collectors are used in gas-cleaning installation systems for simultaneous cooling and moistening (air-conditioning) of gases more often. In this case, they expect function of gas-cleaning installations to carry out a role of heat exchangers of mixture in which the chilled gas stream directly contacts with a chilling liquid (more often water).

The purpose of this book is the analysis of ways of an intensification of process of clearing of gas emissions in the following types of wet gas cleaners: dynamic spray scrubbers, wet gas-cleaning installations of impact-sluggish act, and bubbling dedusters. Experimental research studies and optimization regime and design data of the gas-cleaning installations, which are patented by the Russian Federation, are discussed.

For the sake of everyone, this book is written in the aspect of engineering-technological plot under the circuit design:

- Theoretical bases of process of clearing of gas emissions
- Trial-and-error methods and calculation of apparatuses of wet clearing of gas emissions
- Building of new gas-cleaning installations, their performance characteristics, and recommendations about application.

To bring the attention of the general reader to the book, the newest data under the theory and practice of wet clearing of industrial gases from dispersion particles are systematized. State-of-the-art capability processes of separation of gas-dispersed impurity are observed.

This book can be referred to by students, postgraduates, scientists, and engineers who are engaged in purification of industrial gases.

—— **Regina Usmanova**
Gennady Zaikov

PART I
DYNAMIC SCRUBBERS

CHAPTER 1

MODERN METHODS OF INTENSIFICATION AND DUST CLEARING EFFICIENCY RAISE

CONTENTS

1.1   INTRODUCTION

The rapid development of the industry, which has embraced in the second half of the twentieth century many countries of the world, has led now to a serious decline of ecological situation. One of the burning issues is pollution of air basin by gas emissions of the industrial factories. Growth of industrial outputs has served as the reason of increase in volumes of emissions in a circumambient. Working out of a considerable quantity of new processes promoted increase in quantity of the toxic substances arriving in an aerosphere. The problem of protection of a circumambient can be solved at the expense of a heading of the without waste, self-containedproduction engineering. However, now this direction yet has not had sufficient development; therefore, the problem of creation of the perfect and effective equipment for clearing of gas emissions of the industrial factories is actual[1–14].

The problem essentially becomes complicated that volumes of gas emissions of the industrial factories make tens, and sometimes and hundreds, thousand cube meter per hour that does inconvenient application of the traditional clearing equipment. The majority of the apparatuses used now for clearing of gases from gaseous, liquid and firm impurity, are characterized by the low carrying capacity caused in the small maximum permissible speeds of gas in apparatuses. It serves as the reason of that high efficiency apparatuses have the big overall dimensions (e.g. diameter of tower absorbers can attain 10–12), and expenses for their manufacturing, installation, and transportation are unreasonably great. Besides, in apparatuses of the big diameter, it is impossible to achieve a liquid-phase uniform distribution on their crosssection that leads to sharp decrease in efficiency of clearing[15–20].

The specified problems have served as the reason of that many industrial gas emissions at all are not exposed to clearing. As an example, it is possible to result smoke gases of the factories of metallurgy, power engineering, chemical, petrochemical and other industries, tank and scavenging gases of the various factories, emissions of a dust and steams of organic dissolvents in the production areas of the factories[21–24].

The problem of clearing of great volumes of gas emissions of the industrial factories in an aerosphere can be solved for the account of application for these purposes of apparatuses of whirlwind type. Use in whirlwind apparatuses of centrifugal separation of phases removes restriction on maximum permissible speed of gas and allows to spend processes at the flow rate speeds of gas attaining 20–30 m/s. High carrying capacity of whirlwind apparatuses causes their low metal consumption, rather small specific power expenses, simplicity of manufacturing. Design features of whirlwind apparatuses allow to spend to them complex clearing of gas emissions of the industrial factories as from harmful gaseous impurity, and of small liquid and firm corpuscles. Apparatuses also are rather convenient for conducting in them of process of vapor cooling of high-temperature gas emissions at a stage of preparation of gases to clearing[25–39].

In spite of the fact that a principle of a design of apparatuses of whirlwind type are developed for a long time, their wide use in the industry is restrained by an insufficient level of scrutiny hydro- and aerodynamic regularity of work and absence of reliable and well-founded methods of calculation of efficiency of processes of clearing of gas.

1.2 CONDITION AND PROSPECTS OF DEVELOPMENT OF THE EQUIPMENT OF CLEARING OF GAS EMISSIONS

The intensification of processes leads to a constant decline of an ecological situation in industrially developed centers. It is called by increasing volumes of the gas emissions containing toxic components. The problem of working out of the highly effective dedusters possessing operate reliability, the big carrying capacity, and small power consumption is actual.

The solution of a problem of clearing of the large-scale gas emissions of the factories chemical, petrochemical, and industry allied industries, and also the heat power installations which are burning coal and fuel oil, essentially becomes complicated that traditionally applied equipment for clearing of gas emissions because of its low carrying capacity on gas cannot be used in case of great volumes of gas emissions.

Apparatuses filters (fabric, fibrous, paper, the granulous, etc.), separation in which occurs owing to cogging and inertial intercoupling of corpuscles with a filter medium, under certain conditions can provide high extent of trapping enough fine dust. However, such apparatuses have a high water resistance, are counted for small speed of a filtering and low concentration of a dispersoid, their maintenance demands periodic regeneration of a filter medium. Besides, a heading of fabric filters is often restrained by the restricted sampling of heat-resistant and chemically resistant filtrating cloths. Fabric filters often fail while in service outdoor (Aliev, 1988; 1986).

Wet clearing also is applied to thin and highly effective clearing of gases. Process proceeds at interaction of a stream or vials of gas with a film or liquid drops.

Efficiency of a dust separation depends on the dustiness of gas, sizes of trapped corpuscles, and also from speed of a gas stream and the specific charge of a liquid.

Demanded separation efficiency can be attained at various relationships of speed of gas and a specific irrigation (Kuznetsov, 1989; Kouzov, 1993; and Istomin, 1996). As a rule, choose such regimes of conducting process at which the specific charge of a liquid is minimum.

Wet clearing apply when moistening and gas cooling is admissible, and, corpuscles of a dust separated from gas do not represent worth. Wet clearing should be spent under condition of creation of the closed cycle of circulating water. It is expedient to use this method of clearing of gas emissions when dry ways on that or other condition are inapplicable.

Wet trapping of a dust as a result of contact of corpuscles of a dust with a liquid is carried out in the next ways (Istomin, 1996; and Vatin, 2003):

1. The dusty gas stream arrives in the apparatus and is washed out by a liquid inducted into it or hits about its surface; dust corpuscles leave from a gas stream owing to their collision with liquid drops. Scrubbers refer to the given group of wet-type collectors hollow and with a nozzle, high-speed turbulent dedusters, scrubbers of a percussion and others, for example.
2. Dedusters with wetted surfaces; in them the liquid irrigates a surface of the apparatus or its elements (nozzle) being inside to which the dusty gas stream adjoins. Dust corpuscles are entrained by a film of a liquid and inferred from a gas stream. By this principle wet cyclone separators, scrubbers with a nozzle work, for example.
3. The dusty gas stream is inducted into a liquid and atomized on vials in which dust corpuscles are concluded. Bubbling and foamy apparatuses refer to this group of wet-type collectors.

Dust removal devices can work at a combination of several ways in one apparatus.

To number of deficiencies of wet-type collectors refer to: losses of a liquid owing to a carryover; a decline of conditions of dispersion in an aerosphere of the wet cleared gases, especially aggressive components;

necessity of machining and removal of a considerable quantity of flows and sludge; the big expenses of energy; necessity of application of antirust constructive materials for equipment manufacturing.

Deficiencies of wet-type collectors are compensated: working out of highly effective apparatuses with rather low power consumption, increase in extent of separation of gas and fluid-flow phases, use of a low-purity irrigating liquid for circulation,and so on.

1.3 METHODS OF THE INTENSIFICATION AND RAISE OF EFFICIENCY OF CLEARING OF GAS EMISSIONS

A wide heading of new powerful industrial assemblies with the intensified processes in most cases involves sharp increase in gas emissions. Because of it, loading increases by air-cleaning constructions. There is a problem of an intensification of processes of clearing of gases.

Scientifically, the technical and patent literature (Shkatov, 1981; Shtokman, 1998; and Timonin, 2003)of the considerable quantity of the apparatuses using various methods of an intensification of processes of clearing of gases is presented. Air-conditioning of gases, regime intensification and special ways refer to the last (use of effect of condensation, application is superficial-activesubstances, etc.).

Has gained extending air-conditioning of gases in scrubbers of full transpiration which in the core place in front of electrostatic precipitators, is much more rare—before bag hoses (Belov, 1991).

The constructive-technological intensification of processes of clearing of gases is connected with perfection of a configuration of constructive elements of dust removal apparatuses (Ivanov, 1998; and Kuznetsov, 1989). For example, at cyclone separators improve cases, upstream ends, the overhead covers, exhaust tubes,and sp pm.

As to wet methods of clearing (Timonin, 2006) here, first of all, are exposed to constructive perfection of the device for liquid spraying, and in wider aspect—devices on which character of contact of cleared gas with a liquid depends. So, addition of a design of the foamy apparatus with (Aliev, 1988) stabilizer of foam has allowed in ones two times to raise speed of gas in the apparatus without damage to separation efficiency and without infringement of structure of a foam blanket.

Raise of efficiency of a dust separation is attained in various ways (Rodions, 1989; Rodionov, 1985; and Vatin, 2003). First, demanded intensity of separation is attained at the expense of application of the combined (multistage) circuit designs of clearing of gas emissions; secondly, by use of the high-speed apparatuses working in various hydrodynamic regimes; and, at last, thirdly, use of physical and chemical processing methods of dust-laden streams (clustering, application of surface-active agents, change of wetting of walls of the apparatus, etc.).

The way applied in the whirlwind mass-transfer apparatus with a screw torch of an irrigation, as presented inFigure 1.1 (Aliev, 1988), can be one of the ways of maintenance of an optimum regime of wet clearing of gas emissions.

The gas, inducted it is tangential through a connecting pipe 4, rises on the apparatus the twirled stream. For irrigation of a gas stream by a liquid the deflection sprinkler—the central pipe 2 on which the liquid arrives serves. In a pipe along a spiral plate holes through which the liquid gets out in the form of a stream are executed, hitting about an edge of a plate 6 and forming thus extending fluid-flow film.

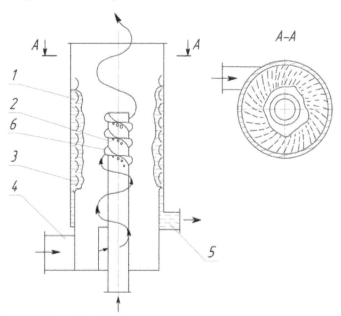

FIGURE 1.1 The whirlwind mass-transfer apparatus with a screw torch of an irrigation:1—the outdoor cylinder; 2—the central pipe; 3—the internal cylinder; 4—a connecting pipe; 5—the point; 6—a spiral plate.

The twirled gas stream destroys a film on small drops which are propelled to a wall of the apparatus under the influence of a centrifugal force and penetrate into a positive allowance between internal 3 and external 1 cylinders through apertures in the internal cylinder. The liquid under the influence of gravitational forces is hauled down in space between internal 3 and external 1 cylinders and leaves through the point 5. The gas separated from a liquid, is inferred from the apparatus in the flue or an aerosphere. Commercial tests have shown a small water resistance of such apparatus.

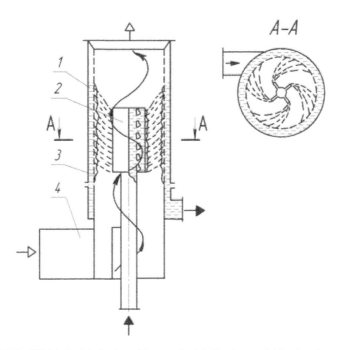

FIGURE 1.2   Whirlwind deduster with a vertical deflection sprinkler:1—the cylindrical body; 2—a sprinkler; 3—a separator; 4—the upstream end.

The whirlwind apparatus for the clearings of gas emissions, presented according to Figure 1.2, on a design and a work principle is analogous to Figure 1.1 whirlwind apparatus shown on Figure 1.3 for absorption of gases (Stark, 1990).

However, for the big overlapping of a gas stream by a fluid-flow film, the baffle of a sprinkler 2 is executed in the form of the plates which have been welded on a pipe along its axis. Gas through the tangential upstream end 4 arrives in the cylindrical body 1, forming a twirled stream. A liquid, getting out holes of a sprinkler 2 in the form of a stream, spreads on a baffle (plate) and forms vertical film curtains on all altitudes of a contact zone. The liquid separates from a twirled gas stream the perforated separator 3. Efficiency of clearing dusty gas from corpuscles a size more than 5 µm depending on diameter of the apparatus makes 90 percent.

But, it is necessary to have in view of that the water resistance reaches 1,500 Pascal. Corpuscles less than 5 µm are not separated by a size.

Besides, in such apparatuses was possibly formation of adjournment in the equipment and gas pipelines at drop ablation of a moisture from a deduster, and also a corrosive wear at clearing of the gases containing aggressive components.

Separator—the wet washer in drawing 1.12 incorporates the vertical cylindrical body 1. In the body the conic swirler 2 containing the silenced lower base 3 and ring upper base with flanging 4 is installed. The Upper base 4 is connected rigidly to the device body. The bases 3 and 4 swirler 2 are rigidly connected with each other by means of unidirectional guide vanes 5. The lower base 3 is supplied by a package of the radial zigzag elements 6 with the apexes oriented in a direction of blade twist of guide vanes 5. Before a swirler 2 the connecting pipe 7 inputs of the liquid, executed with the overhead flat 8 which 3 swirlers 2 rest against a lower base is placed.

The separator—the wet washer works as follows:

The gas stream containing mechanical or gaseous impurity arrives in the device from below. The liquid arrives in the device by means of an axial connecting pipe 7. Thanks to that the connecting pipe 7 inputs of a liquid is placed before a swirler, at liquid supply the zone of contact of phases increases and, hence, the effective utilization of a swept volume of the device occurs more.

The lower base 3 swirlers 2 is executed in the form of a package of the radial zigzag elements 6 against which 8 connecting pipe 7 inputs of a liquid rests a flat overhead shearing. Thus there is a liquid flow on a lower base 3 and formation of flat radial streams. Because zigzag elements 6 partially recoat each other in the plot, the radial fluid-flow streams recoat each other on all crosssection of the device, forming fluid-flow of swirler. The ascending gas stream passes fluid-flow a swirler and gets a rotary motion characterized by intensive turbulent pulsations in a flow core.

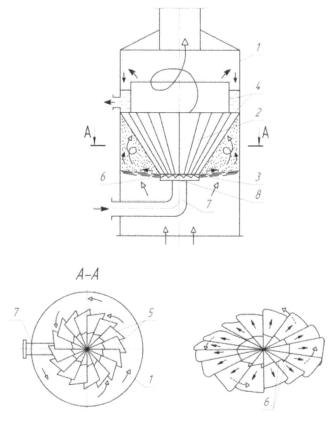

FIGURE 1.3 Separator—the wet washer:1—the body; 2—the centrifugal injector; 3, 4—basis; 5—the guide vane; 6—zigzag elements;7—a connecting pipe; 8—an inclined surface.

Thus, in the twirled stream the liquid, thanks to reflexion of the radial streams of a fluid-flow air swirler from the body of 1 device that causes high-intensity contact of phases outside of a conic air swirler 2 arrives, raises separation efficiency of gas and allows to use a device swept volume more rationally. Further twirled gas–liquid the stream cooperates with guide vanes 5 air swirlers 2 that leads to liquid crushing, updating of a surface of contact of phases and an intensification of processes of an interphase exchange. Thanks to that an apex of zigzag elements 6 are oriented in a direction of blade twist of guide vanes 5, there is an increase in extent of blade twist of a gas–liquid stream after passage of guide vanes 5 that raises efficiency of the subsequent centrifugal sedimentation of a liquid on the body 1. The liquid precipitated on the body 1, drains off in a zone of the ring basis with flanging 4 air swirlers 2, whence is inferred from the device, and cleared of impurity and a dropping liquid gas is venting from the device in the overhead part (Shkatov, 1981).

The condensational deduster presented in Figure 1.4refers to techniques of clearing of gas emissions and allows to raise efficiency of trapping of finely dispersed corpuscles of a dust. The gaseousdust mix with temperature 120–200 C on a bringing tangential connecting pipe arrives in the spiral channel which has been had in a cylindrical part of the body where it twists, moistened with the sprayed water and at the expense of a centrifugal force is bounced to a wall of a jacket of cooling 9 in which on a connecting pipe cold water arrives and on a connecting pipe the completed is taken away. The tubes 10 informed with a cavity of a jacket of cooling 9, are executed perforated, and the refrigerant (water) from the cooling jacket 9 arrives in the channel where it is dispersed to gas streams on the microfogs moistening corpuscles of a dust.

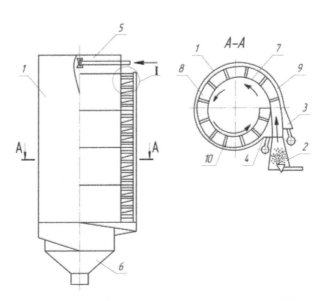

FIGURE 1.4 The condensational deduster:1—the body; 2—a tangential connecting pipe; 3—water feeding; 4—a water leading-out; 5—the exhaust tube; 6—the sludge remover; 7—dividing plate; 8—the channel; 9—a cooling jacket; 10—tubes; 11—a launder.

In the channel, there is a condensation of streams, a stream powdergas. And drops of dispersive water together with dust corpuscles flow off a formed condensate on channel walls partially in the sludge collector, and partially in a launder 11, had under tubes 10. A launder 11 are inclined toward a channel midwall; therefore, water flows off to this wall and the powdergas is dispersed by a stream.

Drops of dispersive water at the expense of a centrifugal force are bounced to an outdoor wall of the channel. There is a repeated circulation of water, its heating, and additional saturation a stream powdergas water steams. The cleared gas stream is thrown out in an aerosphere through the exhaust tube (Shtokman, 1998).

For the purpose of increase in the surface of contact of phases the device is developed for the quality of contact of the phases, which is presented in Figure 1.5.

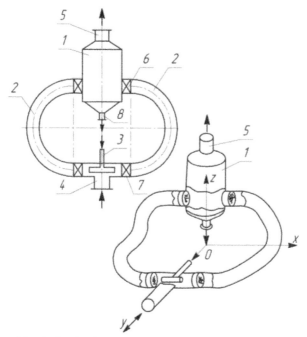

FIGURE 1.5 The device for the quality of contact of phases: 1—the coagulation tank; 2—breakdown pipes; 3—feeding into of an irrigating liquid; 4—gas feeding into; 5—a gas leading-out; 6, 7—air swirlers; 8—an exit branch.

The device contains the coagulation tank 1, breakdown pipes 2 for supply of dusty gas which are bent in a plane or space. The device contains a connecting pipe of feeding into of an irrigating liquid 3, a connecting pipe of feeding into of dusty gas 4, a connecting pipe of tap of the cleared gas 5, the air swirlers 6 installed on target sections of breakdown pipes 2 in front of coagulation tank 1, the air swirlers 7 installed on target sections of breakdown pipes 2, an exit branch 8.

The device works as follows:

Dusty gas through a connecting pipe 4 arrives in breakdown pipes 2, where simultaneously through a connecting pipe 3 submit irrigating liquid. By means of an air swirler 7, installed on an entry in breakdown pipes 2, gas–liquid streams twist in an opposite direction. Because breakdown pipes are bent in a plane or space, that is, executed in the form of curvilinear channels, gas–liquid streams move on a difficult space trajectory, testing act of two fields: fields of the centrifugal force, a created air swirler 7, and fields of a centrifugal force of breakdown pipes 2.

As a result of superposition of acts of this field of a corpuscle of dusty gas and an irrigating liquid, being propelled on a tangent to the curvilinear channel of breakdown pipes, gain a zigzag direction, and in different sections of the channel they have the path, that is, there is their repeated mutual intersection. At the expense of it the surface of contact of phases increases and there is a process intensification. Having attained air swirlers 6, on target sections of the pipes, the prepared gas–liquid streams which have gained high extent of turbulence, are uncoiled by these air swirlers in an opposite side and arrive in the coagulation tank 1 simultaneously from two opposite sides. The cleared gas venting through a connecting pipe 5, and the trapped corpuscles of a dust together with a liquid—through a connecting pipe 8 [37].

The wet-type collector, in drawing Figure1.6 is shown. It consists of the vertical cylindrical body, connecting pipes of 2 and 3 entry and a gas make, a connecting pipe 4 entries of water with spraying injectors 5, a connecting pipe of removal of sludge 6, the sludge remover 7, oscillators of turbulence 8. The turbulence oscillator consists of a fastening lattice 9, springs 10 for a suspension bracket of sheets 11, axis of 12 and distant springs 13.

The deduster works as follows:

The gas dust-laden flow on a connecting pipe 2 arrives through a zone of an irrigation and fastening lattice 9 with big live cross-section in the oscillator of turbulence 8. In the turbulence oscillator together with a gas stream dispersive water (liquid) arrives. Thanks to small distance between sheets 11, equal 5—15 mm, are formed narrow right-angled channels in which at gas high speeds (more than 10 m/s) originate fine-scale three-dimensional pulsations of a gas stream and firm corpuscles being in it and a liquid drop that leads to intensive colliding of corpuscles with liquid drops and a flowing off film of a liquid and by that promotes effective trapping of finely divided corpuscles. Besides, the force of pulsations of a gas stream and a liquid film on all altitude of channels is attained at the expense of the vibration of thin-walled sheets 11 originating at high speeds of gas owing to an insignificant thickness of leaves. Depending on productivity of a deduster the thickness of sheets of metal is equal 0,4—0,8 mm, the thickness of sheets from plastic is slightly more.

The effect of vibration of sheets owing to a gas high speed gains in strength nonrigid fastening of sheets in the turbulence oscillator, that is, as a suspension bracket on springs of 10 sheets making the oscillator, and installation of distant springs 13 between sheets. At an intensive turbulization of a gas stream and a flowing off film of a liquid on all altitude of sheets of oscillators firm corpuscles are trapped not only drops of a dispersive liquid, but also a flowing off turbulence film. The low-purity liquid gets to the sludge remover 7, whence sludge through a connecting pipe 6 is venting from a deduster, and the cleared gas is venting through a connecting pipe 3.

Thanks to direct-flow traffic of the phases, allowing to move gas with a great speed from top to down on the right-angled channels of simple form, in a deduster there are no settling zones that does impossible an overgrowing of internal elements of a deduster by a dust. Besides, owing to simplicity of channels the turbulence oscillator possesses an insignificant water resistance that reduces power expenses for pumping of a gas stream [18].

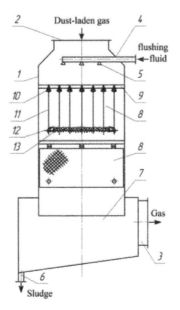

FIGURE 1.6   The wet-type collector:1—the body; 2—a gas entry; 3—a gas make; 4—a water entry; 5—the injector;6—sludge removal; 7—the sludge remover; 8—the turbulence oscillator;9—a lattice; 10—a spring; 11—sheets; 12—an axis; 13—distant springs.

The device for clearing of the gases, presented inFigure 1.7, refers to techniques of wet clearing of gases from dust and allows to trap more full corpuscles of a dust and a moisture drop.

Dusty gas arrives in the body 1 on the tangential upstream end 2. In an entrance zone radially had injectors 3 irrigate the twirled stream with opening splashes.An irrigation promotes concretion of corpuscles and their sedimentation on midwalls of the device and in the bottom part of the apparatus. Before receipt in a connecting pipe 4 cleared gases the dust-laden gas stream passes a zone of guide vanes 5 where smoothly reverses direction with rotational on the forward. On a section of a bend of shovels, there is an additional sedimentation of drops of sludge on inertia and their accent about a surface of shovels. At running off of drops of water on guide vanes 5 there is their merge in streams, which move in launders 6. Obviously expressed landform of launders prevents a mechanical carryover of liquid from their surfaces. The gaseous-dust stream hits vertically in a surface of a trapped liquid and deposits the corpuscles of a dust weighed in it and drops at the expense of sharp veering of a stream [16].

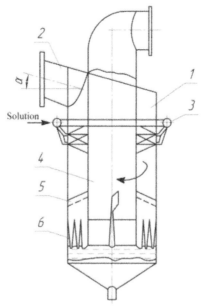

FIGURE 1.7   The device for clearing of gases:1—the body; 2—the upstream end; 3—injectors; 4—an inferring connecting pipe;5—shovels; 6—a launder.

In Figure 1.8,the apparatus executed in the form of the cylindrical body with the tangential upstream end in its bottom part, by a connecting pipe for an exit of the cleared gas—in the overhead part, a connecting pipe for a sludge leading-out is presented. The basic lattice with a nozzle layer over which the spraying device is installed, is executed in the form of the perforated conic surface, a having front-to-back slope, equal to an angle of repose of a used nozzle. The lattice is supplied a louvre separator by a drip pan fixed on a stock under it with possibility of an upright conveyance on a stock, in the overhead part of the body the loading batcher of a nozzle is installed.

The invention raises efficiency of clearing of dusty gases because the apparatusrepresents the combined system, having two step of clearing, first—the cyclone separator with a water film and the another—a gas bottle [16].

All the above-described designs of apparatuses for wet clearing have rather high overall performance, but thus there is a necessity of the organization of a turnaround cycle of water supply that considerably increases the cost of system of clearing of gas emissions.

Besides, the water resistance of such apparatuses attains 400 Pascal as they demand the big water discharge and very much a fog spray accompanied by considerable expenses of energy. In the observed designs was possibly infringement of aerodynamics of air streams at a non-uniform water concentration, the big carryover of liquid.

Thus, the analysis of the observed designs of whirlwind apparatuses allows to draw a conclusion about possibility of working out of highly effective irrigated whirlwind apparatuses for clearing and cooling of gas streams.

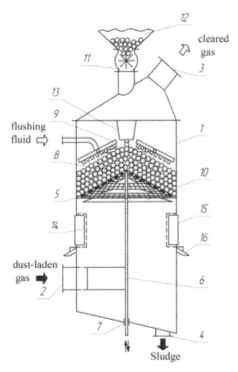

FIGURE 1.8 The apparatus for wet clearing of gases: 1—the body; 2, 3, 4—connecting pipes; 5—a basic lattice; 6—a stock; 7—a support; 8—a nozzle; 9—a sprinkler; 10—a drip pan; 11—the batcher; 12—the loading pocket; 13—a funnel; 14—a window; 15—a cover; 16—tray-type.

Working out of designs a vortex apparatus will allow to raise the efficiency of a dust separation at considerable decrease in water resistance of the device, and also to save material means and the squares of production areas at the expense of possibility of installation of apparatuses in flues of dust removal system.

1.3.1 CONCLUSION

The analysis of merits and demerits of various methods of clearing gases from gaseous impurity has allowed to draw a conclusion that at clearing of great volumes of gas emissions by the most simple in implementation the method of centrifugal separation is reliable.

However, traditionally applied separation equipment possesses low carrying capacity on gas. In this connection, use of dynamic spray scrubbers is in certain cases the unique way of the solution of a problem.

Comparison of possibilities of the equipment for clearing of gas of various types also testifies to preference of application of such gas bottles at clearing of great volumes of gas emissions. However, it is necessary to note that for the further raise of efficiency of gas bottles conducting theoretical and experimental researches is required. Experiment will allow to installgasdynamic circumstances and to define the most rational designs.

1.4 PROBLEMS OF CALCULATION OF HYDRODYNAMICS AND SEPARATION

Creation of universal mathematical model of the turbulent twirled stream is a necessary condition for working out of adequate methods of calculation of gas bottles of different technological appointment, their optimization regime and design data, abbreviation of expensive experimental researches. Now in connection with rapid development of computer production engineering essential progress was outlined in working out of more and more difficult mathematical model approaches to calculation of gas bottles.

Generally such model should secure, on the one hand, with as much as possible exact forecasting of parameters of clearing of gas emissions at change of this or that essential parameter, on the other hand, reception of the information on possibly ways of an intensification of process of separation. To ensure this data, the model should give the information on all momentous aspects of a current (geometrical characteristics, boundary conditions, physical properties of medium, turbulence, etc.), and to secure with possibility of the solution of the equations. The mathematical models, which are coming nearer to a reality, share both by the way of increase in dimensions of a quantity of model, and by the way of the exact description of proceeding physical processes. Difficulties originating thus are connected with two aspects: modeling and the solution. As it is underlined in Ref. (Kuznetsov, 1989), construction of models now is far from end even monophase turbulent currents though variety enough effective models is already known, and calculation of many of them does not call basic complexities. For the description of hydrodynamics of low concentrated heterogeneous streams where agency of a dispersoid on traffic of the bearing medium is insignificant, such models appear quite comprehensible.

Hydrodynamic models of gas bottles can be divided on two groups— the models of perfect fluid based on the solution of Euler equations, and viscous fluid models in which basis are assumed Navier–Stokes equation or Reynolds. The analysis of models of a nonviscous stream (Goldshtik, 1981) allows to conclude that the assumption of usability to gas bottles of the circuit design of perfect fluid in the core justifies; however, observed models of a potential, screw, or general whirling motion do not reflect variety of the features generated by turbulence and viscosity of medium (such, as formation of an axial circulating zone, sticking of medium in firm walls, etc.)

More adequate results give models of a whirl of a viscous fluid (Uzhov, 1981). However, here again there is variety of the difficulties connected with short circuit of the initial equations, consistent statement of boundary conditions, bulkiness of gained models, and complexity of their numerical implementation the successes attained in this direction, allow to observe it as more research planning.

Modeling of dynamics of a dispersoid in a gas bottle completed question first of all character traffic of polydispersion particles in the turbulent twirled stream, the complicated interacting of corpuscles with walls, with each other and with a bearing turbulent stream. Difficulties are called by indeterminate form of entrance conditions, possibly change of number and a size of corpuscles as a result of crushing or concretion, a deformation of corpuscles (drops, bubbles), and also other reasons.

Plain models of dynamics of a dispersoid in apparatuses can be realized within the limits of Lagrangian field theory, assuming integration the equation of traffic along separate paths of corpuscles (Timonin, 2006). More difficult approach is based on free motion of phases (Baranov, 1989) when the equations of traffic and energy for both phases register in Eulerian coordinates and dare on uniform algorithm. In calculation of gas bottles while it is a little instance of use of the continuous approximation. Its application is represented to the admissible to the quick-response corpuscles which sizes there is less than scale of turbulence of a carrying stream.

The determined description of traffic of corpuscles in the turbulent twirled stream is based on the solution of the equations only for average values, without interacting with casual fields of pulsations of speed of a continuous phase. According to Ref. (Vatin, 2003) application of the given approach it is represented justified for the corpuscles which size is commensurable with scale of turbulence or exceeds it. The determined approach is well formalized and, despite the simplified nature, possesses certain advantages, in particular, at satisfactory accuracy allows to avoid the difficulties originating at integration of the equations of intermittent traffic of corpuscles. Besides, it is more universal, as it eliminates necessity of empirical approximation of parameters of type of factor of a turbulent diffusion or factor of intensity of the casual affectings used in diffusion and a stochastic model (Mikhaylenko, 2008).

1.5 PROBLEMS OF DESIGNING OF SCRUBBERS

Numerous publications concerning practical application of scrubbers give the grounds to ascertain that the area of industrial extending of scrubbers steadily extends every year. Use of scrubbers allows to raise compactness and individual productivity of installations for gas clearing, to scale down a payback period, to raise efficiency of proceeding processes. Efficiency of scrubbers is noted by authors of all mentioned above monographies and reviews on these subjects.

However, there are factors complicating spreading mud process of scrubbers. Despite observed rough expansion of scopes of wet clearing of gas, there is no data about full replacement of equipment with them (cyclone separators, wet filters, scrubbers, etc.) at least in one of the industries (Aliev, 1986; Belov, 1991; and Kolesnik, 1986).

Principal causes of restricted use of scrubbers many authors (Ivanov, 1998; Usmanova, 2008; and Vatin, 2003) consider absence of reliable methods of calculation of aerodynamics and separation processes in scrubbers and criteria of transition from laboratory models to the large-scale installations.

According to the data (Baranov, 1989; and Mikhaylenko, 2008), the additional constraint is unstable stability of work of the scrubbers, manifested in essential change of parameters of work at infinitesimal changes of entrance conditions. In particular, the basic deficiency of hydrocyclonic apparatuses is defined by considerable change of parameters of separation at small oscillations of concentration and composition of the firm phase in apparatus.

In Reference (Vatin, 2003; and Rodionov, 1985), it is noted that now versatile application of the twirled streams advance process of their research. It leads to thereto that there are many individual highly effective gas-cleaning installations, but their wide circulation is restrained by absence of accurate recommendations for transition to other productivity or change of operating modes. Analogous remarks can be found and in other survey research studies on these subjects.

The desire to intensify processes in whirlwind apparatuses leads to necessity of essential increase in a twisting of a stream. With a strong twisting stability loss is peculiar to streams that is expressed in disintegration of an axisymmetric whirlwind kernel and origination after a zone of disintegration of several types of the indignant traffic. According to the data (Ivanov, 1998), the site of disintegration of a whirlwind and possibility of its emersion are defined by values of a Reynolds number and twisting parameter, and the increase in a twisting of a stream leads to earlier origination of disintegration, that is, expands boundary lines of unsteady operation of whirlwind apparatuses.

In-process (Ivanov, 1998), it is offered to conduct complex research of work for each design of a scrubber. Thus, to make a card of an operating mode which would reflect boundary lines of resistant to existence of stationary whirlwind structures and boundary line of recommended areas of maintenance of apparatuses After that was possibly to realize the offer (Istomin, 1996; and Uzhov, 1981) about raise of stability of work of apparatuses at the expense of application of systems of automatic control.

Concerning a current state of exploratory works under the theory and practice a gas-cleaning installation, it is necessary to notice the following. Under the data [203] only on apparatuses of wet clearing for the 10-year-old period in the world scientific and technical literature has appeared more than 2,500 scientific publications. The analogous result has given and the analysis of patent researches presented there. Approximately the same situation is observed and in other spheres of application of scrubbers. Such wide scope of researches speaks first of all seeming simplicity of implementation in plants of the twirled streams listed above-mentioned"useful" properties.

To execute complete analysis of all published materials it is not obviously possible. However, in the light of modern representations to state an estimation of existing problems in observed area, the problem quite real. It is connected with that till now in a huge stream of publications the works ignoring difficult structure of a real current predominate. Many experimental research studies still are restricted only to measurement of mean field characteristics of a stream on which it is impossible to gain representation about the space structure of a real current.

Without research of three-dimensional structure of sharp twirled stream, it is impossible to size up correctly new design solutions and to find boundary lines of conversion zones of work for whirlwind apparatuses, it is impossible to define, whether peak efficiency of target process, and the main thing is attained,—it is impossible to create the similarity theory of sharp twirled streams necessary for realized and purposeful perfection of gas-cleaning installations.

In the absence of such theory in the literature, the considerable quantity separated (at times inconsistent) recommendations for choice design data of scrubbers has by this time collected that the turn it has generated an unreasonable variety of industrially released apparatuses with various twisting devices and setting sizes. Offered recommendations, as a rule, are based only on experimental data that reduces a range of their usability. By way of illustration (in Table 1.1) the generalized intervals of recommended values from the big selection of such recommendations collected in-process (Ivanov, 1998; and Rozengart, 1985). Latitude of dispersion of recommended values are resulted serves as the additional certificate of trouble of a current situation.

TABLE 1.1   The recommendation for sampling of design data

Parameter	View	Value
Relative width of the upstream end	b/D	$0.05 \div 0.35$
Relative diameter of the upstream end	d/D	$0.15 \div 0.75$
Relationship of sizes of the upstream end	h/b	$1 \div 6$
Relationship of the square of an entry and exit	A_{in}/A_{ot}	$0.6 \div 2.5$
Relative length of the exhaust tube	l_{in}/D	$0.5 \div 1.8$
Relative length of the body of the apparatus	l/D	$1.5 \div 5.5$
The relation of a conic part to length of the body	L_k/l	$0 \div 1$

The problem of development of the uniform approach to designing of scrubbers is aggravated with disunity of exploratory groups and their affiliation to various industries.

It is enough to tell that till now in a scope of gas-cleaning installations there is no uniform nomenclature, are not developed uniform specifications size series of apparatuses and there is no general approach to an estimation technological, parameters of apparatuses. The analogous situation is observed also industrial production of scrubbers, and most the acute problem of serial manufacturing costs concerning dynamic spray scrubbers for which in connection with their small gabarits the manufacturing methods from sheet metal rolling widely applied in case of the cyclonic equipment is unacceptable.

By data (Rodions, 1989; and Rodionov, 1985), till now in Russia dynamic spray scrubbers are produced in the conditions of job-lot production (it is maintenance shops and mechanical department of the factories more often). In world practice manufacture of dynamic spray scrubbers also is not centralized

1.6   CONCLUSIONS

By results of the spent analysis it is possible to conclude that owing to the obvious operational advantages dynamic spray scrubbers have found application in the diversified industrial production engineering. Conservation of trends to expansion of scopes of dynamic spray scrubbers continues to compel attention researchers and designers in Russia and abroad.

At the same time, processes in scrubbers remain till now little-studied. Now on hydrodynamics and separation of nonuniform systems in scrubbers the extensive experimental material, however the gained results is saved up restrict to measurement of mean field characteristics of a stream and integrated parameters of separation on which it is impossible to gain representation about the space formation of a real current, about features of dynamics of a dispersoid and about possibly ways of an intensification of separation processes.

Not studied there are questions on change of operating characteristics of apparatuses at change of regimes with various whirlwind formations. Designs of apparatuses for which would be spent full research of operating modes are not known yet. In creation of methods of calculation of scrubbers the simplified models ignoring difficult formation of a real current and reducing thereby accuracy of forecasting of technological parameters of apparatuses till now predominate.

Effect of the form of corpuscles of a dispersoid on efficiency of clearing of gas emissions is poorly studied. This point in question has basic value for separation of nonuniform systems with not isometric or astable-dispersoid.

In the field of designing of scrubbers there is no uniform methodology, there are heterogeneous size series of designs with excessively wide intervals of a variation of geometric proportions of the setting. There are no strictly well-founded generalized recommendations about rational sampling of proportions of the basic constructive elements. At last, there is no centralized serial exhaustion of the technological and effective scrubbers, capable to fulfill a growing demand for them.

In this connection in the given work, following problems are put:
- To carry out the complex structural analysis of hydrodynamics of the bearing medium in a scrubber for the purpose of refinement of representations about real formation of the turbulent twirled stream and revealing of reserves of perfection of the apparatus.
- To study regularity of traffic of dispersion particles and to develop the model of separation process considering effect of the basic constructive sizes on efficiency of separation.
- On assay values of hydrodynamics and separation of dispersion particles to formulate recommendations about designing the scrub-

bers directed on decrease of power consumption of processes of clearing of gas.

- On the basis of the formulated deductions and recommendations to devise gas-cleaning installations for conditions of serial exhaustion and large-scale implementation in industrial practice.

KEYWORDS

- **Clearing of gas emissions**
- **Designing**
- **Hydrodynamics**
- **Separation**
- **The Dynamic spray scrubber**
- **The Industry**

REFERENCES

1. Muhutdinov, R.N; and Artamonov, N.A.; Whirlwind mass transfer the apparatus the copyright certificate 861914 USSR;**1981.**
2. Ladygichev, M. G.; Foreign and domestic equipment for gas purification: a reference book. M: Teplotekhnik;**2004.**
3 Lakomkin's, A.A.;A separator—scrubber. The copyright certificate 1421379 USSR;**1988.**
4. Logachyov, A.P.; and Voronina, E.A.; The condensational deduster. The copyright certificate 1430073 USSR;**1988.**
5. Letjuk, A.I.;Tkach, A.I.; and Gridasov, V.N.; The device for the quality of contact of phases. The copyright certificate1185674 USSR;**1995.**
6. Patent 1438829 USSR, MKI 47/06.The Device for Clearing of Gas.Korotkov, JU.A.;and Tchernikov,
7. Nechaev, J.G.;Yesipov, G.P.; and Rudenko, G.V.; A Wet Deduster.The patent 2091137 Russian Federations.**1997.**
8. Zhuravlyov, V.P.; and Valiev, A.M.; The Apparatus for Wet Clearing of Gases.The patent of 2054306 Russian Federations;**1996.**
9. Aliev, M.;TheDevice and Service Gas-Cleaning Installation and Dust Removal Facilities. Moscow: Metallurgy;**1988.**
10. Aliev, G. M.-Ç.; Technics of a Dust Separation and Clearing of Industrial Gases: The Directory. Moscow: Metallurgy;**1986.**

11. Baranov, D.A.;Kutepov, A.N.; and Lagutkin. M.P.;To calculation of difficult circuit designs of the joint of hydrocyclone separators.*The Appl. Chem.***1989**,*62(11)*,2486–2490, *(in Russian)*.

12. Belov, S.T.; Preservation of the Environment. Eds. Belov, S.T.; Barbinov, F.A.; Kozjakov, A.F.;Moscow: The Higher school;**1991.**

13. Belevitsky, A.M.; Designing Gas-Cleaning Installation. L.: Chemistry;**1990.**

14. Bespamjatnov, G. P.; Maximum Permissible Concentration of Substances in Surrounding Medium. L: Chemistry;**1985.**

15. Clearing and a RegenerationofPlant Emissions./Ed.Maksimova, V. F.;Moscow: Wood Prom; **34(12), 1981.**

16. Evremov, G. I.; and Lukachevsky, V.P.; Clearing of Gas Emissions.Moscow: Chemistry;**1990.**

17. Goldshtik, M. A.; Vortex Flows. Novosibirsk: The Science;**1981.**

18. Ivanov, A.A.; Calculation and Designing of Whirlwind Separation Apparatuses on the Basis of the Structural Analysis of Hydrodynamics of the Twirled Streams. Dissertation Assoc. Prof. Dr. Ing. Dzerzhinsk;**1998,** 307 p, *(in Russian)*.

19. Kuznetsov, I.E.; Equipment for Sanitary Cleaning Gases. Kiev:Tehnika;**1989.**

20. Kouzov, P. A.;Malgin, A.D.; and Skryabin, G.M.; Ochistka of Gases and Air from a Dust in the Chemical Industry. St. Petersburg: Chemistry;**1993.**

21. Mikhaylenko, G.G.; New Atomizing Device in Industrial Gas Treatment: Monograph. Odessa: Astroprint;**2008.**

22. Kolesnik, A.A.; and Nikolaev, N.A.; Experience in testing and commissioning of dedusting installations in oil refineries and petrochemical plants. M: Petrochemistry;**1986.**

23. Rodions, А.И.; Technics of Protection of a Circumambient. Moscow: Chemistry;**1989.**

24. Rodionov, A.I.; Smiths, J.P.; and Zenkov, V.V.;Oborudovanie, Constructions, Bases of Designing of Himiko-Technological Processes of Protection of Biosphere from Plant Emissions. Moscow: Chemistry;**1985.**

25. Rodionov, A.I.;Klushin, V. N;and Sister, V.G.; Processes of Ecological Safety. Kaluga:N.Bochkarevs Publishing House; **2000.**

26. Rozengart, J.I.; Heat Power Engineering of Metal Works.Rozengart, J. I.;Muradova, Z. A.;Teverovsky, B. Z.; et al. Moscow: Metallurgy;**1985.**

27. Technician Manual; Recommendations about Designingof Air Purification from a Dust in Exhaust Ventilation Systems. Moscow: Building Publishing House;**1985.**

28. Stark, S.B.; Gas-Cleaning Installations and Installations in Metallurgical Manufacture. Moscow: Metallurgy;**1990.**

29. Shkatov, E.F.; Automation of Industrial and Sanitary Cleaning Gases in Chemical Promyshlennosti.Moscow: Chemistry;**1981.**

30. Shtokman, E.A.; Air Purification. Moscow: Publishing House ASV;**1998**.

31. Istomin, V.A.; Gas Hydraytesa in Russia: Meeting the Challenge.In "Gas in the CIS";**1996.**

32. Timonin, A.S.;The Engineering-Ecological Directory. T. 1-3. Kaluga: N.Bochkarevs Publishing House;**2003.**

33. Didenko, V.G.; Tech Scrubbing Ventilation Emissions: Textbook. Volgograd: VolgGASU;**1996.**

34. Timonin, A.S.; Bas of Designing and Calculation of the Himiko-Technological and Nature Protection Equipment: The Directory. T. 1-3. Kaluga: N.Bochkarevs Publishing House;**2006.**

35. Uzhov, V. N; and Valdberg, A.U.; Clearing of Industrial Gases from a Dust. Moscow: Chemistry;**1981**.

36. Uspensky, V. A.; Theory, Calculation and Researches of Whirlwind Apparatuses of Treatment Facilities—Assoc. Prof. Dr. Ing. Dzerzhinsk Moscow; **1983,** (*in Russian*).

37. Usmanova,R.R.;Panov, A.K.; and Zaikov, G.E.; Hydrodynamic and Mass Transfer in Vortical-Type Devices. Nova Science Publishers;**2008.**

38. Vatin, N.I.; and Strelets, K.I.; Air Purification by Means of Apparatuses of Type the Cyclone Separator. St. Petersburg;**2003.**

39. Zajcik, L.I.; and Pershukov, V.A.;Problemy of Modelling Gas-Liquid Turbulent Flows with Changes of Phase.*Mech. Liquid Gas.***1996,***5*, 3–19, (*in Russian*).

CHAPTER 2

NUMERICAL SIMULATION AND CALCULATION OF DISTRIBUTION OF THE FLOW RATE OF GAS IN THE APPARATUS

CONTENTS

2.1   INTRODUCTION

From the very beginning of emersion of scrubbers before engineers who designed them, there was a problem—to predict parameters of work of the machine created by them before drawings will be given to manufacture. For apparatuses of rotational act, the problem became complicated that critical parameters of their work (resistance, efficiency, an input, etc.) are caused by a pattern of a current of streams in the setting. The current of multiphase medium so difficult that quite often unique reliable method of research in hydrodynamics remains till now anexperiment. Only in the past years, essential progress in creation of simulars and calculation of currents of multiphase medium has been attained [1-10]. It has given the chance to carry out calculation with high reliability of gained results that the necessary volume of experiment in many cases is reduced to a minimum. Unlike experiment, the numerical approach gives the chance to vary a row of the important parameters of a problem, such as viscosity, angular speed of twirl of the model, essentially influencing formation and behavior of the convective currents [11-18].

The purposes of the given work:
1.   To develop algorithm of modeling of process of separation of firm corpuscles in a dynamic scrubber, allowing to define potential possibilities of apparatuses of gas cleaning.
2.   To carry out the analysis of the numerical scheme to build numerical model in a packet of computing hydrodynamics *Ansys CFX*. To reveal the presence of whirling motions and gain distribution of speed and pressure of a dust-laden gas stream.
3.   To execute numerical research of a flow pattern of a current of a stream in programmable complex *Ansys CFX*. To compare the results of calculations for various flows. To make verification of the data gained by calculation with the data gained in experiment.

2.2   SURVEY OF MATHEMATICAL MODELS OF MULTIPHASE CURRENTS

Conditionally discriminate following types of multiphase currents. The first case—the observed volume completely is filled by substance of one phase (e.g. liquids), and the substance of other phase meets in this volume in the form of discrete corpuscles (a firm phase) or vials (gas phase), and the volume fraction of substance of other phase is insignificant (to 10% of total amount). The second case—the observed volume is partially filled by a liquid, and partially—gas which do not mix up among themselves and are separated from each other by a free surface. The third case—the most difficult—substances of various phases can mix up among themselves (to dissolve/to deposit from a solution), and the volume fraction of substance of other phase is great (over 10% of total amount). For modeling of these types of multiphase currents, various approaches, which we now and will observe, are used [15-22].

2.2.1 MODEL OF DISPERSION PARTICLES

This approach is used for modeling of two-phase currents to which the substance of one of phases is presented in the form of dispersion particles, and the volume fraction occupied with these corpuscles, is insignificant (to 10% of total amount). Instances of such currents are spluttered in a stream of air of a drip of water, air bells in a fluid stream, and also form corpuscles in air or water stream. The substance forming a master phase is necessary continuous medium, and its current is modeled by the equations of Nave—Stokes (1) (or Reynolds (3)) and continuity (2). The substance which is present in a stream in the form of discrete corpuscles does not form the continuous medium, separate corpuscles cooperate with a stream of a master phase and is with each other discrete. For modeling of traffic of corpuscles of the absent-minded phase, the approach of Lagranzha is used, that is, traffic of separately taken corpuscles of the absent-minded phase under the influence of forces from outside a master phase stream is traced.

Problems such quite often meet in the chemical industry, see Ref. (Kochevsky, 2004).

For simplicity, assume that corpuscles of the absent-minded phase have the spherical form. The forces acting on this corpuscle are caused by a difference of speed of a corpuscle and speed of a stream of a master phase, and also replacement by this corpuscle of medium of a master phase. The equation of traffic of such corpuscle has been inferred in-process (Kochevsky, 2003) and written as

$$m_p \frac{dv_p}{dt} = 3\pi\mu d C_{cor} \left(v_f - v_p\right) + \frac{\pi d^3 \rho_f}{6} \frac{dv_f}{dt} + \frac{\pi d^3 \rho_f}{12} \left(\frac{dv_f}{dt} - \frac{dv_p}{dt}\right) +$$

$$+ F_e - \frac{\pi d^3}{6}\left(\rho_p - \rho_f\right)\vec{\omega}\times\left(\vec{\omega}\times\vec{r}\right) - \frac{\pi d^3 \rho_p}{3}\left(\vec{\omega}\times v_p\right). \qquad (2.1)$$

Here, m_p is the weight of the corpuscle; d is the diameter of the corpuscle; υ is the speed; μ is the dynamic viscosity of substance of a master phase; C_{cor} is its factor of viscous resistance; $\acute{\omega}$ is the angular speed of twirl; r is the radius of a vector. The index p (particle) refers to a corpuscle, and index f (fluid) is for a substance of a master phase.

The left-hand side of Eq.(2.1) represents the sum of all forces acting on a corpuscle, expressed through weight and acceleration of this corpuscle. The first member on the right-hand side expresses a retardation of a corpuscle as a result of a viscous friction about a master phase stream by the law Stokes. The second member is the force enclosed to a corpuscle, owing to a differential head in a master phase surrounding a corpuscle, the stream of a master phase called by acceleration. The third member is the force demanded for acceleration weight of a master phase in volume, expeled by a corpuscle. These two members are necessary to consider when the master phase density surpasses the density of corpuscles, for example, by consideration of vials of air in a fluid stream. The fourth member (F_e) is a superposed force directly acting on a corpuscle, for example, gravity or force of electric field. Last two members—centrifugal force and force of Koriolisa occur only by traffic consideration in a relative frame of reference. Besides, sometimes in the right-hand side (15), it is necessary to consider some additional forces (e.g. in the presence in a stream of considerable difference of temperatures).

Eq.(2.1) represents a differential first-order equation in which unique unknown magnitude is speed of a corpuscle v_p, and argument—a time t. Speed of substance of a master phase v_f in all space points is necessary known. In the capacity of initial data, except the size and properties of a corpuscle, its rule during the initial moment of a time is set. It is underlined also that should occur at collision of a corpuscle with a wall or with other corpuscle. For performance of calculation the members containing v_p are transferred to the left-hand side of Eq.(2.1). Speed and a corpuscle position during each subsequent moment of a time is defined by a numerical integration on a time with some step Δt with all other members of Eq.(2.1).

The factor of viscous resistance C_{cor} at a moderate Reynolds number $0.01 < Re_p < 1000$ can be computed by formula that follows

$$\tilde{N}_{cor} = \begin{cases} 1 + 0.1315 \left(\text{Re}_\rho \right)^{0,82-0,05\alpha} & \text{Re}_\rho < 20 \\ 1 + 0.1935 \left(\text{Re}_\rho \right)^{0,6305} & \text{Re}_\rho > 20 \end{cases} \quad \text{при}$$

где $Re_p = \rho_f \mid v_f - v_p \mid d \, / \, \mu, \quad \alpha = \log Re_p$.

There is also a possibility to model heat—and mass transfer between dispersion particles and main stream available in modern software products. For example, transpiration liquid drops at low enough pressure of gas in a surrounding stream or in enough heat. The algorithms realized in *Ansys CFX* allow to model affecting on a stream of substance of a master phase and moving in it of discrete corpuscles. As a first approximation, the density and viscosity of substance of a master phase are multiplied with $(1-\alpha_p)$, where α_p is the specific volume occupied with discrete corpuscles. Further on each step on a time changes of weight, pulse and energy of discrete corpuscles are computed. These changes are added accordingly in the equations of conservation of mass, a pulse, and energy for a master phase stream. Thus, calculation of a current of a master phase and traffic of dispersion particles is carried out in common.

If the stream of substance of a master phase is turbulent, the mechanical trajectory of dispersion particles is not determined as it depends on intensity and a direction of turbulent pulsations. The way of modeling of affecting the turbulent pulsations of a main stream on traffic of dispersion particles has been offered, in particular, in-process (Aksenov, 1996).

In modern software products, some boundary conditions matching to various events, occurring are realized at a collision of a discrete corpuscle with a firm wall: a recoil as a result of elastic or not elastic blow, sticking to a wall, slippage along a wall (depending on physical properties and a collision angle), passage through a wall (if a wall porous), and another. There is also a possibility of modeling of splitting and merge under certain conditions droplets of water or gas vials at their collision with each other.

2.2.2 MODEL OF CURRENTS WITH A FREE SURFACE

Thegiven approach allows to model a current of two (or more) liquids or liquid and gas which do not mix up with each other and, being in the field of mass forces, form among themselves an accurate boundary, that is, a free surface.

According to the given approach, the mathematical model for free surface approximation is supplemented with a transport equation of function of filling F expressing "concentration of a liquid in gas" (by consideration of a current of a liquid with gas). The name of model of a current—model VOF (the volume occupied with a liquid) from here as follows:

$$\frac{\partial F}{\partial t} + \frac{\partial}{\partial x_j}\left(Fpu_j\right) = 0 \qquad (2.2)$$

In the area occupied with a liquid, it is $F = 1$, in the area occupied with gas, it is $F = 0$. Only in meshes through which there passes a free surface, $0 < F < 1$. In the capacity of the entry condition, the free surface initial point is set.

The algorithm of numerical calculation at use of such model is presented in Refs (Lampart, 2001; and Launder, 1974).

2.2.3 MULTIPHASE MODEL OF MIXTURE (MULTIPHASE MIXTURE MODEL)

The given approach allows to model a current of multiphase medium that can mix up among themselves, and do not form a free surface.

For modeling of a current of two or several phases in this model one continuity equation, one set of the equations of traffic, and one equation of the energy, written concerning ensemble averaged on weight of values of speed and mix density is used. So, the continuity equation in this model looks like:

$$\frac{\partial \rho_m}{\partial t} + \frac{\partial}{\partial x_j}\left(\rho_m u_{mj}\right) = m \qquad (2.3)$$

where ρ_m is the mix density, u_m is the ensemble averaged speed on weight, u_{mj} is the a projection of speed to an axis x_j. m, by default equal to null, represents a mass transport owing to cavitation and other physical effects.

The given model allows to consider that traffic of various phases occurs to various speeds, using the concept of speeds of sliding. It allows to model, for example, a retardation of a stream of the grains of sand flying in the reservoir, filled with a motionless liquid.

The traffic equation in a projection to an axis x_1 in this model looks like:

$$\frac{\partial}{\partial t}\left(\rho_m u_{mi}\right)+\frac{\partial}{\partial x_j}\left(\rho_m u_{mi} u_{mj}\right)=-\frac{\partial \rho}{\partial x_i}+\frac{\partial}{\partial x_j}\left[\mu_m\left(\frac{\partial u_{mi}}{\partial x_j}+\frac{\partial u_{mj}}{\partial x_i}\right)\right]+f_i+\frac{\partial}{\partial x_i}\left(u_{ki}-u_{mi}\right)^2, \quad (2.4)$$

where μ_m is the ensemble averaged viscosity on weight, u_k is the speed of substance k-й a secondary phase, u_{mj} is the projection of this speed to an axis x_1, (u_k-u_m)—speed of slippage of substance k-й a secondary phase concerning ensemble averaged speed on weight u_m. Eq. (2.4) differs from Eq. (2.1) by the presence of last member modeling mutual slippage of phases.

The energy equation is similarly written.

For separately taken k-й, a secondary phase it is possible to present a continuity equation in an aspect

$$\frac{\partial\left(F_k \rho_k\right)}{\partial t}+\frac{\partial}{\partial x_j}\left(F_k \rho_k u_{mj}+F_k \rho_k u_{kj}\right)=0, \quad (2.5)$$

where ρ_k is the substance density k-й phases. From this equation, it is possible to define volume fraction F_k occupied with substance k-й of a phase in some mesh of space.

2.2.4 EULER'S MULTIPHASE MODEL

The given model is most general and the most difficult among models of a multiphase current. The substance in each of the phases is necessary continuous medium, and traffic of substance of each of phases is modeled by own system of the equations of Nave—Stokes (Reynolds), continuity, and energy.

According to this model, the equations of traffic written for each phase, dare in common. The algorithm of calculation of such currents has been offered, in particular, in-process (Menter, 1994), and realized in *Ansys CFX*. The given model is the most exacting to computing resources of the computer—and to an online storage size, and to speed of the processor.

2.2.5 ADDITIONAL RECOMMENDATIONS FOR CHOICE MODELS OF A MULTIPHASE CURRENT

For modeling of currents in which substances of various phases can mix up and do not form a free surface, in many cases it is possible to use both model of dispersion particles, and mixture model, and Euler's multiphase model. Additional criteria of sampling of appropriate model such are as follows (Menter, 2001).

- The relation β weights of substance of a dispersoid (d) to weight of substance of a bearing phase (c):

$$\beta = \gamma \frac{F_d}{F_c},$$ (2.6)

F_d and F_c are volume fractions, γ is the relation of density of a disperse and bearing phase, $\gamma = \rho_d / \rho_c$; this relation is equal to 1000 for firm corpuscles in a gas stream, is equal 1 for firm corpuscles in a fluid stream, and is equal 0.001 for gas corpuscles in a fluid stream.

At very low relation β, dispersion particles practically do not influence a stream of a bearing phase, and it is possible to use any of the listed models. At very high values β, dispersion particles strongly influence a stream of a bearing phase, and for appropriate modeling of a current it is necessary to use only Euler's multiphase model. At a mean β, for sampling of suitable model, it is necessary to compute Stokes number as it is presented more low.

- Number of Stokes St:

$$St = \frac{t_d}{t_c},$$ (2.7)

where t_d is the time characterizing traffic of corpuscles, $t_d = (\rho_d \, d_d^2)/(18 \, \mu_c)$, d_d is the diameter of a corpuscle, μ_c — is the viscosity of substance of a bearing phase, $t_c = L_c/U_c$ is the time characterizing a current of the bearing phase, L_c is the characteristic length, and U_c is the characteristic speed.

At $St \ll 1,0$, dispersoid corpuscles do not deviate almost

From streamlines of a bearing phase, also it is possible to use any model of a current (as a rule, mixture model—with the least resource). At $St >$ 1.0, the dispersoid particle path does not coincide at all with streamlines of a bearing phase, and the mixture model in this case is unsuitable: it is necessary to use either model of dispersion particles or Euler's multiphase model.

Modeling in packet *Ansys CFX* for calculation of hydrogas kinetics of a scrubber consists of five stages:

1. Creation of $3D$ models
2. Creation of a desing grid
3. A statement of problem (preprocessor)
4. The solution of mathematical model
5. Definition of magnitudes and evident visualization (postprocessor)

2.3   GEOMETRICAL MODEL CREATION

The first stage of preparation of initial data for current calculation is creation of the solid-state geometrical model simulating volume in which there is an investigated current. As elements of the setting of a dynamic scrubber (the twisting device, inlet branches, and entrainment separators) often have rather difficult form, creation of their solid-state model is not trivial problem. $3D$ apparatus model has been executed in a packet of solid-state modeling *Solid Works* (Figure 2.1 see), and then imported in *Ansys Design Modeler.*

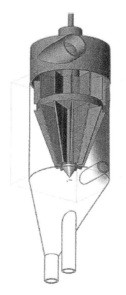

FIGURE 2.1   Geometrical model of a scrubber in *solid works.*

2.4   CONSTRUCTION OF THE DESING GRID

Quality gained on the basis of conducting of computing experiment of results directly depends on the quality of the built desing grid. Preprocessor *GAMBIT* allows to create and process sweepingly geometry of investigated processes. *Ansys Mesh* possesses the powerful oscillator of the grids, allowing to create various types of grids: the structured hexahedral grid, automatic (not structured) hexahedral and a grid tetrahedron. Besides, in it there is a possibility of creation of boundary layers with the combined grids. After construction of a grid, the user has possibility to muster its quality on various parameters (displacement of elements, a relationship of sides).

Construction of a desing grid is the process of division of desing area on assemblage of separate meshes. Grid meshes represent polyhedrons, usually tetrahedrons, hexahedrons, prisms, or pyramids (Figure 2.2). Edges of these meshes form lines of a desing grid, and the points which have been had on edges or in the center of meshes are knots of a desing grid. As a result of the numerical solution of the equations of mathematical model in knots of a desing grid required parameters of a current are also defined (Wilcox, 1986).

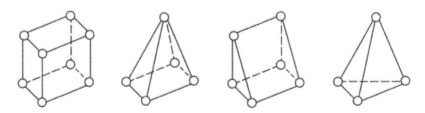

Hexahedron Pyramids Prism Tetrahedron

FIGURE 2.2   Typical forms of meshes of a grid.

The basic demand to a desing grid—it should be enough dense to allow the physical effects occurring in desing area. For achievement of uniform accuracy of calculation, grid knots should place more densely in places of sweeping change of parameters of a current, in particular at walls. Besides, at grid construction, it is necessary to avoid reception of excessively drawn down or warped meshes which form too strongly differs from correct polyhedronsin the presence of such meshes can it is essential reception of the converging solution (Aksenov, 1996) will be at a loss.

Discriminate the structured and unstructured desing grids. In unstructured desing grids, grid knots are scattered in space in a random way according to the set law of density of an arrangement of knots. It does possible construction of a grid in area of as much as difficult form. However, difference analogs of the equations of mathematical model on such grid appear bulky. For construction of the structured grid, the desing area breaks into blocks according to some topology of breakdown set by the user, and within each block the desing grid to which knots it is possible to refer under numbers of a three-dimensional file is under construction. Application of such grid allows to organize the most economic algorithms of calculation (Crowe, 1998).

In our problem, the right-angled, adaptive, locally comminuted is final-volume grid in-process was applied to the solution of the equations of mathematical model.

The grid used for the solution of the given problem is presented in drawings (Figures 2.3 and 2.4 see).

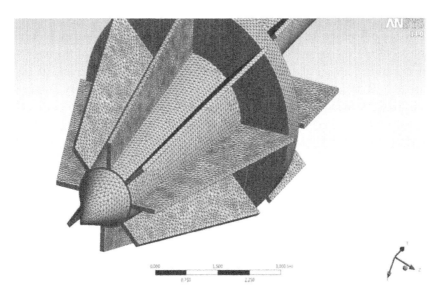

FIGURE 2.3   Typical desing area, a desing grid, and a surface of the interface of a twirled vortex generator.

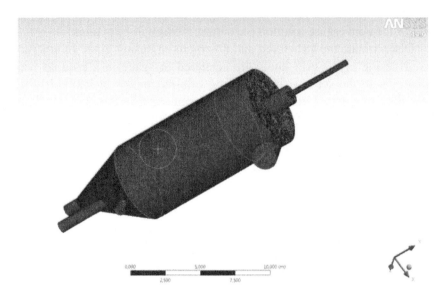

FIGURE 2.4   Typical desing area, a desing grid, and an interface surface in a dynamic scrubber.

Sizes of one mesh of an initial grid make 3×10^{-3} mm. Adaptation of first level is lead on surfaces: a conic asymmetrical surface of the case of the apparatus, a wall of a tangential connecting pipe of feeding into of gas, entries of an axial flow of a liquid, and a peripheral flow of slurry. Adaptation of second level is lead on surfaces of a twirled vortex generator, in areas of blade passages. Desing components have been set from database *Ansys CFX*: air, pure water, and a disperse dust. Numerical calculation in the given program complex does not allow to model a torch of the sprayed liquid and to size up its dispersity as in the program the account of a surface tension force of a liquid and its interaction with a circumambient is not put. Therefore, result of numerical experiment are the gained values of speed of gas and liquid streams and pressure in any point of the crank chamber which allow to investigate in details process of interacting of streams, to choose an optimum relationship of sizes of a scrubber and to define a range of its work (Vasquez, 2000).

The final stage in grid construction is the boundary layer task. In the given problem, the boundary layer consists of five parts, and the minimum mesh width in a boundary layer makes 0.1 mm.

As a result quantity of meshes of the generated grid has made more for the apparatus of an order of 2 million meshes, for a vortex generator of an order of 1,00,000 meshes (see Figure 2.5).

Table 1. Mesh Information for CFX

Domain	Nodes	Elements
Default Domain	390941	2065909
rotat	114048	91200
All Domains	504989	2157109

FIGURE 2.5   Sizes of a desing grid.

In preprocessor *Ansys CFX*, it is necessary to execute a statement of problem which includes the following moments:
- Working medium sampling
- The task of entry conditions for calculation
- Sampling of mathematical model
- The task of boundary conditions
- The task of parameters of the solution

2.4.1   WORKING MEDIUM SAMPLING

In a dynamic scrubber in the capacity of operating fluid, water viscosity $\mu_{ж} = 1 \cdot 10^3$ Ns/m² was used; factor of a superficial tension of a liquid $\sigma = 72,5 \cdot 10^3$ N/m; dusty gas with $\rho_r = 1,291$ density of dusty gas, kg/m³; $\mu_r = 0,0000189$—viscosity of gas, the Pas·s; v—kinematic viscosity of gas, $15,56 \cdot 10^6$ m²/s; for a corpuscle of a quartz dust ($\rho = 2600$ kg/m³) diameter $d=1 \div 150$ a micron, moving in air stream ($V = 5,1 \div 35 \cdot 10^{-6}$ m²/c). In entry conditions for calculation, magnitude of an ambient pressure and ambient temperature is set.

Magnitude of external (surrounding) pressure makes 1 atm; temperature of ambient air is 25°C.

It is necessary that on lateral walls the attachment condition is satisfied.

$$U_{wall} = 0,$$

On upper bound and in the field of a flow, values of speed are set:

$$
\begin{cases}
U_{inlet} = u_{axial}\bar{i} + u_{radial}\bar{j} + u_{swirl}\bar{k} \\
u_{axial} = -u_1 \\
u_{radial} = 0 \\
u_{swirl} = 0
\end{cases}
,
\begin{cases}
U_{inlet} = u_{axial}\bar{i} + u_{radial}\bar{j} + u_{swirl}\bar{k} \\
u_{axial} = -u_2 \\
u_{radial} = 0 \\
u_{swirl} = 0
\end{cases}
\tag{2.8}
$$

Investigated problems dare in axisymmetric statement (dependence on azimuthal coordinate φ is not considered), and the liquid current is supposed to be turbulent and is presented by the system of the operating equations in the dimensional formulation.

2.4.2 SAMPLING OF MATHEMATICAL MODEL

The mathematical model of traffic of gas in the apparatus is based on the solution of system of the equations of Nave–Stokes for an axisymmetric problem and a continuity equation (Bache, 2001).

$$\frac{1}{r}\left[\frac{\partial}{\partial r}(r\rho v_r v_r) + \frac{\partial}{\partial z}(r\rho v_z v_r)\right] = \frac{1}{r}\left[\frac{\partial}{\partial r}\left(r\mu_T \frac{\partial v_r}{\partial r}\right) + \frac{\partial}{\partial z}\left(r\mu_T \frac{\partial v r}{\partial z}\right)\right] - \frac{\partial P}{\partial r} - \mu_T \frac{\rho v_r}{r^2} + \frac{\rho v^2 \phi}{r}$$

$$\frac{1}{r}\left[\frac{\partial}{\partial r}(r\rho v_r v_\phi) + \frac{\partial}{\partial z}(r\rho v_z v_\phi)\right] = \frac{1}{r}\left[\frac{\partial}{\partial r}\left(r\mu_T \frac{\partial v_\phi}{\partial r}\right) + \frac{\partial}{\partial z}\left(r\mu_T \frac{\partial v_\phi}{\partial z}\right)\right] - \mu_T \frac{\rho v_\phi}{r^2} - \frac{\rho v_\phi v_r}{r}$$

$$\frac{1}{r}\left[\frac{\partial}{\partial r}(r\rho v_r v_\phi) + \frac{\partial}{\partial z}(r\rho v_z v_\phi)\right] = \frac{1}{r}\left[\frac{\partial}{\partial r}\left(r\mu_T \frac{\partial v_\phi}{\partial r}\right) + \frac{\partial}{\partial z}\left(r\mu_T \frac{\partial v_\phi}{\partial z}\right)\right] - \mu_T \frac{\rho v_\phi}{r^2} - \frac{\rho v_\phi v_r}{r} \tag{2.9}$$

$$\frac{1}{r}\left[\frac{\partial}{\partial r}(r\rho v_r v_z) + \frac{\partial}{\partial z}(r\rho v_z v_z)\right] = \frac{1}{r}\left[\frac{\partial}{\partial r}\left(r\mu_T \frac{\partial v_z}{\partial r}\right) + \frac{\partial}{\partial z}\left(r\mu_T \frac{\partial v_z}{\partial z}\right)\right] - \frac{\partial P}{\partial z}$$

$$div\ \rho\bar{\upsilon} = 0 \tag{2.10}$$

where υ_z is the speed of a stream along an axis; υ_g is the speed of a stream in the radial direction; υ_ϕ is the tangential speed of a stream; p is the mix density; μ is the factor of turbulent viscosity; P is the pressure; υ is the vector of speed.

For short circuit of system of the equations the two-parametric model of turbulence to—ε, as one of the most well-proved models for calculation of currents, such is used. At turbulence modeling, it was used $to\varepsilon$ model, for it dares two additional transport equations for the purpose of definition to turbulent kinetic energy and εturbulent energy of a dissipation.

Model turbulence:

$$\frac{\partial \rho_\hbar k_c}{\partial t}\alpha_c + \frac{\partial \rho_c u_c k_c}{\partial x_j}\alpha_c = \tau_{ij}\frac{\partial u_c}{\partial x_j}\alpha_c - \alpha_c \cdot \rho_c \cdot \varepsilon_c + \frac{\partial}{\partial x_j}\left[\alpha_c\left(\mu_c + \frac{\mu_c'}{\sigma_k}\right)\frac{\partial k_c}{\partial x_j}\right]$$

$$\frac{\partial \rho_\hbar \varepsilon_c}{\partial t}\alpha_c + \frac{\partial \rho_c u_c \varepsilon_c}{\partial x_j}\alpha_c = C_{\varepsilon 1}\frac{\varepsilon_c}{k_c}\tau_{ij}\frac{\partial u_c}{\partial x_j} - C_{\varepsilon 1}\frac{\varepsilon_c^2}{k_c} + \frac{\partial}{\partial x_j}\left[\alpha_c\left(\mu_c + \frac{\mu_c'}{\sigma_\varepsilon}\right)\frac{\partial \varepsilon_c}{\partial x_j}\right]$$

$$\mu_c' = C_\mu \cdot \rho_c \cdot \frac{k_c^2}{\varepsilon}. \tag{2.11}$$

where κ_c is the a turbulent kinetic energy of a gas phase; σ_κ is the turbulent Prandtl number for the kinetic energy equation; μ_c and μ are molecular and turbulent viscosities of a gas phase; ε_c is the speed of a dissipation of a turbulent kinetic energy; σ_ε is *the* turbulent Prandtl number for the equation of a dissipation of a kinetic energy; τ_{ij} is the Cartesian components of tensor of voltage: $\mu = 0.09, \varepsilon_1 = 1.44, \varepsilon_2 = 1.92, \sigma_k = 1.0, \sigma_\varepsilon = 1.3$.

Calculations show that near to firm walls there is a peracute change of parameters k and ε. For the appropriate permission of these changes, it is necessary to use rather dense desing grid.

2.5 DIGITIZATION OF THE EQUATIONS OF MATHEMATICAL MODEL

As it is known, the basic approaches to a digitization of the equations of mathematical model are MFD (finite difference method), MFE (a method of final elements), and MFV (a method of final volumes). All of them can be observed as variety of the more general approach named a method of weighed discrepancies. MKP is, perhaps, most simple and intuitively clear; however, its application is at a loss on not structured grids. MFE equally successfully works on the structured and not structured grids that does convenient its application in areas of as much as difficult geometrical configuration. The important advantage of MFV is maintenance of con-servation relations of integrated magnitudes (the charge, a momentum) on each of meshes of a desing grid, and not just in a limit, in process of enough strong thickening of a desing grid.

In *Ansys CFX*, MFV is used with elements of the final-element ap-proach that allows to combine the specified advantages of these methods.

2.6 ALGORITHM OF THE NUMERICAL SOLUTION OF THE EQUATIONS OF MATHEMATICAL MODEL

2.6.1 THE JOINT SOLUTION OF CONTINUITY EQUATIONS AND TRAFFIC

According to the algorithm offered in Crowe, 1998 and realized in *Ansys CFX*, the numerical solution of the Eqs.(2.9) and (2.10) within one global iteration is carried out not consistently, and in common. It leads to repeated increase in a size of a matrix of Slough, complication of its structure and algorithm of the solution of Slough and accordingly to increase in a time of calculation at one global iteration. Nevertheless such approach justifies itself owing to essential increase in speed of convergence of algorithm as a whole—for convergence achievement, the smaller number of global iterations is required.

2.6.2 THE TASK OF BOUNDARY CONDITIONS

In the task of boundary conditions, statements of a problem on an exit and an entry in the setting of desing geometry (Figure 2.6 see) are specified. In the capacity of boundary conditions, as a rule, adhesion condition on all firm walls (speed is equal to null), distribution of all components of speed in entrance cross-section and equality to null of the first derivatives (in a current direction) components of speed in target cross-section are set. Following boundary conditions were set: on firm surfaces, the condi-tion of a smooth wall was set, speed and a temperature gradient on a wall were equated to null: on an entrance surface, the general mass flow rate and stream temperature were set: the two-phase stream (continuous and a dispersoid) was set by volume fractions from the general mass flow rate: distribution of corpuscles of a dispersoid in entrance cross section was uniform: on target surfaces the condition on pressure was laid down.

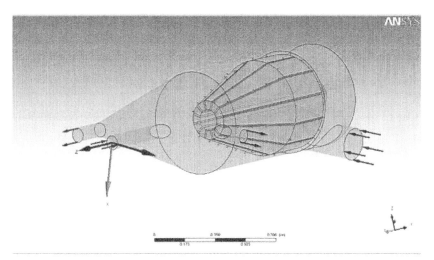

FIGURE 2.6 Computational model boundary conditions.

After the task of boundary conditions, **solution parameters** are set.

Here the convergence criterion (criterion RMS—a mean-square error) is underlined,

The lapse of scalings (by default 10^{-4}) is underlined.

Management of the solution is underlined on global variables that allows to reveal an error of calculations at scaling of the first several iterations. Such errors can originate at the incorrect task of boundary conditions and incorrectly built grid. For the given problem, it is necessary to set quantity of iterations not less than 1,000 iterations (Figure 2.7 see).

Visible	Yes
Transparency	0,1
Color	
Definition	
☐ Suppressed	No
Coordinate System	Default Coordinate System
Reference Frame	Lagrangian
Material	
Fluid/Solid	Defined By Geometry (Fluid)
Bounding Box	
Length X	1,94 m
Length Y	0,6689 m
Length Z	0,69632 m
Properties	
☐ Volume	0,24933 m³
Centroid X	-0,25575 m
Centroid Y	-1,2474e-002 m
Centroid Z	2,371e-004 m
Statistics	
Nodes	390941
Elements	2065909
Mesh Metric	None

FIGURE 2.7   Statistics of a desing grid.

The problem dared in stationary statement, the current was observed as two-phase, effects of turbulence were inducted by means of two-parametric model of turbulence κ–ε.

2.7 VISUALIZATION AND THE ANALYSIS OF RESULTS OF CALCULATION

Process of calculation of a current is carried out before achievement of the set convergence criterion. By default in *Ansys CFX,* results of calculation of a current remain in the form of a file containing all information on a current, sufficient to renew calculation from last condition. It is possible to specify to the program to keep and intermediate conditions that will allow to visualize to a time history of a pattern of a current. Thus, in a file of results co-ordinates of all knots of a desing grid and value of critical parameters of a current are stored in these knots.

The set of available visualization tools *Ansys CFX* usually switches on: and an isosurface (a line and a surface of equal values of some parameter), versicolored pouring, animation of traffic of corpuscles of a liquid, and so on. Desing area with the pattern of visualization put on it is possible to move, increase the two-dimensional schedule, a circuital field, isograms, and so on, and to write this process in the form of an animation film.

In *Ansys CFX* possibility of reception of integrated parameters of calculation, including typical for dedusters is realized also: the hydraulic resistance, a pressure, an input, efficiency of clearing, swirling flow, and is possibility to edit the formula on which these parameters are computed.

2.7.1 RESULTS OF NUMERICAL EXPERIMENT

As a result of numerical experiment static pressures of a gas stream in all cross sections of desing space that has allowed to size up an apparatus hydraulic resistance have been gained. Distribution of static pressure is presented in the form of color on values (Figures 2.8–2.10 see).

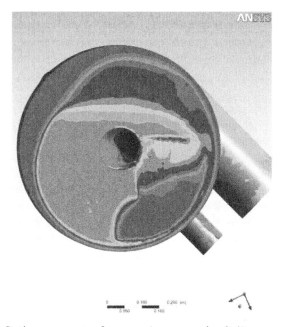

FIGURE 2.8   Static components of pressure (a cross-section 0–1).

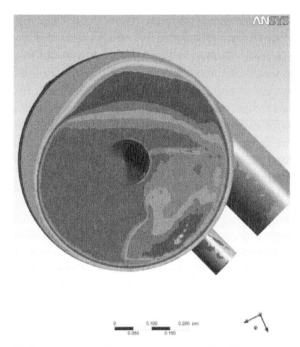

FIGURE 2.9   Static component of pressure (a cross-section 0–2).

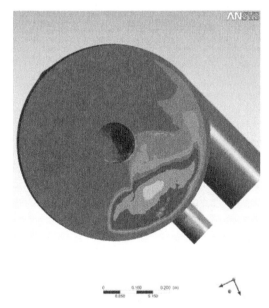

FIGURE 2.10   Static component of pressure (a cross-section 0–3).

It is necessary to pay attention that in eddy zones the underpressure both in comparison with a main stream, and in a zone of blades of a vortex generator is observed. Irregularity of static making pressures in a scrubber has reducing an effect on efficiency of clearing. By comparison to empirical data on separation efficiency it is determined that decrease in efficiency of separation does not exceed 1 percent, though on level of irregularity of a pressure pattern a difference more appreciable. It is possible to explain it to that irregularity of pressure is compensated by positive effect of presence of the eddy zones promoting branch of small corpuscles of a dust from a main stream in zones of a rarefaction and their removal on a spiral path from working space, and further on walls of a conic part of the apparatus in the sludge remover. On apparatus altitude high enough energy of turbulent pulsations remains, it leads to the best dispersion of corpuscles in a stream and to increase in efficiency of separation.

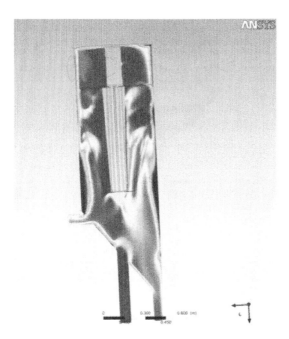

FIGURE 2.11   Static component pressures (longitudinal section).

It is necessary to register that the total pressure in the given program complex develops from superfluous and dynamic, without the atmospheric. The atmospheric pressure is set in reference values, therefore the zero on a scale of pressures (Figure 2.12 see). From drawing (Figure 2.11 see) it is visible that the total pressure of a dust-laden gas stream in the apparatus case increases to periphery. On discharge connections pressure drops to the atmospheric.

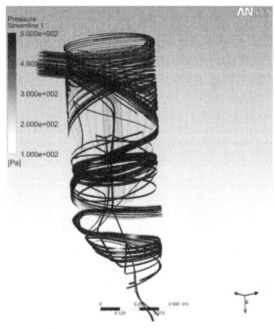

FIGURE 2.12   Pressure in a stream.

As a result of numerical experiment distribution of axial and tangential speed in a longitudinal section of working space of the apparatus (Figures 2.13–2.16 see) is gained. From drawings it is visible that the velocity distribution has the peripheral uniformity, therefore we will observe distribution of axial, tangential and radial components of speed in the form of the stream-lines passing along cross-section of the apparatus.

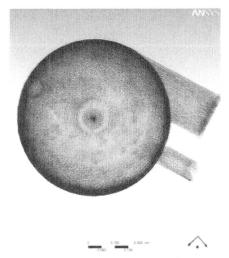

FIGURE 2.13   Projections of a vector speeds (cross-section).

From drawings it is visible that the radial velocity (Figure 2.14 see) keeps constant value practically on all cross-section of working space whereas the axial velocity (Figure 2.15 see) diminishes from the centre to periphery, and tangential, on the contrary, increases (Figure 2.16) see. The gained results it will be co-ordinated with the literary data, and speak about interacting in the apparatus of two streams—forward and rotational.

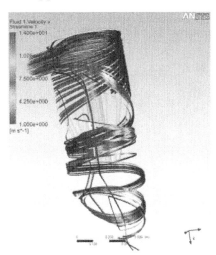

FIGURE 2.14   Speed in a stream v.

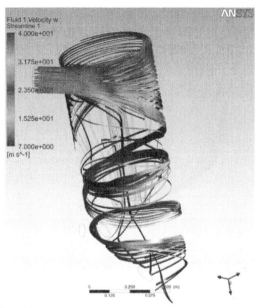

FIGURE 2.15   Speed in a stream w.

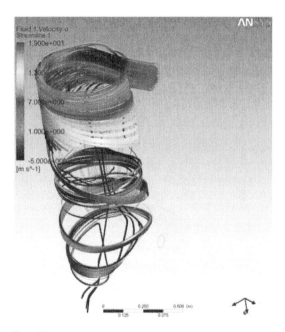

FIGURE 2.16   Speed in a stream u.

The magnitude of tangential speed gained by the numerical solution, qualitatively will be coordinated with experiment. The profile of tangential speed is in the form of parabolas with the maximum which has been had more close to a cylindrical wall that it is possible to explain act of a centrifugal force. Experimental results of the measured speeds of a stream in an axial direction show that near to an exhaust connecting pipe in a wall layer intensive enough current occurs. In the central part of a stream the axial velocity has actually constant value. Numerical modeling by calculation axial a component of speeds of a stream shows the qualitative consent with empirical data.

2.8   OPTIMIZATION

In *Ansys CFX* means of optimization of critical bucklings of desing area (in particular, module Optimus) are also provided (Vasquez, 2000). Thus it is necessary to set the task of optimization, that is, to specify optimization parameters, an admissible range of their change, other restrictions and criterion function, and also to specify an optimization method. On each step of optimization the program completely will execute numerical calculation of a current in working area with a matching set of critical bucklings. Certainly, the calculation time thus appears extremely big that complicates use of such approach. Nevertheless, in the future this approach, probably, becomes very much a designing powerful tool of hydrodynamically perfect gas clean apparatuses.

2.8.1   COMPARISON OF RESULTS ANSYS CFX WITH RESULTS OF EXPERIMENT

In experiment following values have been gained

TABLE 2.1   Results of experimental researches a component of speed

Cross-section 0–1							
V_x.cp.= 15 m/s		V_x.cp.= 19m/s		V_x.cp.= 15 m/s		V_x.cp.= 19 m/s	
Vφ, m/s	Vx, m/s	Vφ, m/s	Vx, m/s	Vφ,m/s	Vx, m/s	Vφ, m/s	Vx, m/s
9.262	5.141	17.016	9.444	12.121	8.184	16.433	14,809

TABLE 2.1   *(Continued)*

Cross-section 0–1

V_x.cp.= 15 m/s		V_x.cp.= 19m/s		V_x.cp.= 15 m/s		V_x.cp.= 19 m/s	
17.078	9.479	25.937	14.396	19.859	8.856	29.389	14.355
23.976	11.971	35.799	18.265	25.608	9.339	38.515	12.914
30.594	13.643	45.212	20.161	30.88	7.721	45.111	11.280
35.149	14.944	50.469	21.458	32.814	6.999	48.382	14.826
36.869	14.549	51.768	20.952	34.416	7.340	50.701	12.677
36.212	12.142	51.36	17.721	36.393	5.142	51.612	8.213
35.903	10.66	50.588	15.021	35.781	3.788	51.072	8.127
35.354	9.830	48.766	13.559	35.025	3.399	49.415	7.863
33.3	8.326	46.15	12.399	33.468	2.366	47.348	5.849
31.052	9.516	44.429	11.936	32.349	3.424	43.653	5.392
26.87	9.271	37.549	12.227	28.504	4.536	41.005	5.065

Cross-section 0-2

V_x.cp.= 8m/s		V_x.cp.= 15 m/s		V_x.cp.= 15 m/s		V_x.cp.= 19 m/s	
Vφ, m/s	Vx, m/s	Vφ, m/s	Vx,m/s	Vφ, m/s	Vx, m/s	Vφ, m/s	Vx, m/s
4.517	6.455	7.353	10.507	14.884	12.760	9.365	15.596
10.083	5.596	13.814	9.327	20.517	11.481	20.414	13.784
17.690	4.423	26.935	7.236	33.282	9.195	37.089	10.661
24.160	3.844	34.528	5.494	42.099	7.340	48.899	9.540
28.737	4.060	37.587	5.645	47.068	6.584	53.971	8.106
29.425	3.115	38.517	4.077	48.231	5.077	56.005	7.912
28.248	0.022	36.512	−0.608	45.877	−0.764	51.455	0.041
27.437	−0.936	34.958	−0.582	44.653	−0.743	49.264	0.039
26.409	−0.901	34.139	−1.164	42.752	−0.712	48.125	−0.801
24.909	−0.415	33.336	0.027	39.38	0.718	44.554	0.813
22.579	0.018	30.182	0.551	36.459	1.301	41.986	1.499

Cross-section0-3

Vx.cp.= 15 m/s		Vx.cp.= 19 m/s		Vx.cp.= 15 m/s		Vx.cp.= 19 m/s	
Vφ, m/s	Vx, m/s	Vφ, m/s	Vx, m/s	Vφ, m/s	Vx, m/s	Vφ, m/s	Vx, m/s

TABLE 2.1 *(Continued)*

Cross-section 0–1							
V_x.cp.= 15 m/s		V_x.cp.= 19m/s		V_x.cp.= 15 m/s		V_x.cp.= 19 m/s	
				0	0	3,398	2,471
4.660	0.329	7.791	0.414	2.632	1.401	6.821	4.434
8.567	0.907	12.453	1.208	6.518	3.617	9.484	6.166
11.559	2.465	16.500	3.219	11.110	5.669	16.266	9.403
14.557	5.309	20.546	7.493			23.398	11.682
18.298	8.938	25.375	12.669	20.035	9.357	29.844	14.577
21.017	12.149	29.631	17.825	24.101	9.268	33.244	12.785
22.839	14.848	32.331	21.831	25.826	9.418	35.690	12.663
22.666	18.371	34.985	24.98	26.365	9.097	35.905	12.039
23.015	19.674	34.845	27.249	26.063	8.739	34.968	10.052

From the data presented in Table2.1, it is visible that values making a component of speed of the stream, gained in CFX and experimentally, are similar. Experimental data will well be co-ordinated with numerical results at values of a specific irrigation to magnitude L/G =0.9 and essentially discriminated at increase in supply of a liquid. This results from the fact that with growth of magnitude of an irrigation the real current investigated in experiment, loses an axial symmetry, and in calculations in a program complex we lay down a condition of an axial symmetry of a stream.

TABLE 2.2 Results of research of a hydraulic resistance

Angular speed of twirl, ω, s^{-1}	Speed of gas on an entry, W,mps	Pressure losses (experimental values)		Design values,ΔP, Pa
		ΔP, mm.water column	ΔP, Pa	
$\overline{D} = 0,927; \overline{l} = 0,072; \alpha = 0^0; z = 12$				
80	2.86	4	39	38
	5.9	8.8	86	74
	10.7	13.6	133	125

TABLE 2.1   *(Continued)*

	15.4	18.4	180	178
	19.7	28.8	282	274
	22.8	39.2	384	376
	25.7	50.4	494	490
	29.4	61.6	604	584
	30.1	72	706	684

$$\bar{D} = 0,927; \bar{l} = 0,172; \alpha_y = 0^0; z = 12$$

150	2.86	4	39	40
	5.9	8.8	86	79
	11.4	18.4	180	172
	16.7	29.6	290	276
	21.5	40.8	400	367
	28.7	50.4	494	460
	31.4	60.8	596	608
	14.8	80	784	784

From the data presented in Table2.2, it is visible, results of numerical modeling will well be co-ordinated with empirical data. Deviations of the gained results do not exceed 7 percent that speaks about adequacy of the offered model.

2.9   CONCLUSIONS

In the given work the algorithm of modeling of process of separation of a dispersoid in a gas stream with irrigation by a liquid has been developed. The carried out calculations allow to define potential possibilities of a dynamic scrubber at its use in the capacity of the apparatus for clearing of gas emissions. Verification of the data gained by calculation, and also an estimation of the parameters defining possibility of separation of a dispersoid on drips of an irrigating liquid, is modeled as process of a current of a water gas stream in a packet of computing hydrodynamics *Ansys CFX.* Numerical research of work of a scrubber will allow to analyze its work for the purpose of decrease of power inputs at conservation of quality of gas cleaning. The developed model helps to simulate traffic of a dusty gas stream sweepingly and visually. The model can consider modification of geometry of the apparatus. Thus, the model can be applied to optimization of a design of a dynamic scrubber.

KEYWORDS

- **Ansys CFX**
- **Components of speed**
- **Desing grid**
- **Numerical modeling**
- **Pressure**
- **Scrubber**

REFERENCES

1. Aksenov, A. A.; Dyadkin, A. A.; and Gudzovsky, A. V.; Numerical simulation of car tire aquaplaning. Computational Fluid Dynamics. John Wiley&Sons;**1996**, 815–820.
2. Bache, G.; CFX-BladeGen Version 4.0 reaches new heights in blade design. CFX Update – Spring. **2001**,*20,* 9.
3. Crowe, C.; Sommerfield, M.; and Yutaka, T.; Multiphase Flows with Droplets and Particles. CRC Press;**1998.**
4. Doormaal, J. F.; New grid generation for rotating machinery. CFX Update – Autumn.**2002**,*22,* 6.
5. Dukowicz, J. K.; A particle-fluid numerical model for liquid sprays.*J.Comput. Phys.***1980**,*35,* 229–253.
6. Grotjans, H.; and Menter, F. R.; Wall functions for general application CFD codes. In: ECCOMAS 98 Proceedings of the Fourth European Computational Fluid Dynamics Conference. John Wiley & Sons;**1998**,1112–1117 p.
7. Hah, C.; Bryans, A. C.;Moussa, Z.; and Tomsho, M. E.; Application of viscous flow computations for the aerodynamic performance of a back*k*swept impeller at various operating conditions.*J.Turbomach.***1988**,*110,*303–311.
8. Harlow, F. H.; and Welch, J. E.; Numerical calculation of time-dependent viscous incompressible flows of fluid with free surface. *Phys. Fluids.* **1965**,*8,* 2182–2187.

9. Hirt, C. W.;and Nicholls, B. D.; Volume of Fluid (VOF) method for dynamical free boundaries.*J. Comput. Phys.* **1981**,*39*, 201–225.

10. Kochevsky, A.N.; Raschet of internal currents of a liquid in channels by means of software product flow vision. *Bull. SumGu.*2004, *2(61)*, 25–36, *(in Russian)*.

11. Kochevsky, A.N.; and Nenja, V.G.; Modern the approach to modelling and calculation of currents of a liquid in bladed hydromachines. *Bull. Sum. Gu.*2003,*13(59)*, 195–210, *(in Russian)*.

12. Lampart, P.;Swirydczuk, J.; Gardzilewicz, A.;Yershov, S.; and Rusanov, A.; The comparison of performance of the menter shear stress transport and baldwin-lomax models with respect to CFD prediction of losses in HP axial turbine stages. *Technol. Fluid/Thermal/Struct./Chem. Sys. Indust. Appl.* ASME. 2001,*1(424)*, 1–12.

13. Launder, B. E.; and Spalding, D. B.; The numerical computation of turbulent flows. *Comp. Meth. Appl. Mech. Eng.* 1974,*3*, 269–289.

14. Menter, F. R.; Multiscale model for turbulent flows in 24th fluid dynamic conference. *Am. Inst. Aeronaut. Astronaut.* 1993.

15. Menter, F. R.; Two-equation eddy-viscosity turbulence models for engineering applications. *AIAA J.* 1994,*32(8)*.

16. Menter, F. R.; and Esch, T.; Advanced Turbulence Modelling in CFX.CFX Update.2001,*20*, 4–5.

17. Patel, V. C.;Rodi, W.; and Scheuerer, G.; Turbulence models for near-wall and low reynolds number flows: a review. *AIAA J.*1985,*23(9)*, 1308–1319.

18. Patankar, S. V.; and Spalding, D. B. A; Calculation procedure for heat, mass and momentum transfer in three-dimensional parabolic flows. *Int. J. Heat Mass Trans.*1972,*15*, 1787–1806.

19. Rusanov, A. B.; and Ershov, S.V.; The Method of Calculation of Three-Dimensional Turbulent Flows in the Setting of any Form. Kharkov: In Problems of Engineering Industry of Ukraine; 2003, 132–136 p.

20. Singhal, A. K.; Li, H. Y.; Athavale, M. M.; and Jiang, Y.; Mathematical Basis and Validation of the Full Cavitation Model. ASME FEDSM'01 – New Orleans, Louisiana;2001.

21. Vasquez, S. A.; and Ivanov, V. A.; A phase coupled method for solving multiphase problems on unstructured meshes. In: Proceedings of ASME FEDSM'00: ASME 2000 Fluids Engineering Division Summer Meeting. Boston;2000.

22. Wilcox, D. C.; Multiscale model for turbulent flows in AIAA 24th aerospace meeting. *Am. Inst. Aero. Astro.*1986.

CHAPTER 3

EXPERIMENTAL RESEARCH STUDIES AND HYDRAULIC RESISTANCE CALCULATION

CONTENTS

3.1   INTRODUCTION

The problem of decrease in gas emissions for the purpose of maintenance of admissible concentration of dust in air basin can be solved, if for each case study legitimately to choose economic and effective enough deduster. The cyclonic clearing of plant emissions passed round now of suspended matters has cost 10 times less than their wet clearing, and also clearing in bag hoses and electrostatic precipitators. However, cyclone separators used under production conditions not always secure with demanded quality of clearing and have other defects (Batting, 2003). To perspective methods of raise of efficiency of a dust separation of finely divided corpuscles, it is possible to refer to wet clearing of gas. For this method, difficult mass transfer processes occurs during the interaction of a gas-dispersed stream with drops of an irrigating liquid are characteristic [1-10]. Therefore, speed and the concentration of phases defining clearing of gas emissions change.

Available research studies in this area show strong sensitivity of output characteristics to a regime and an apparatus build that testifies about qualitative to various hydrodynamics of streams at different values regime design data. Difficult character of velocity distribution, pressure gradient presence on radius in the twirled gas stream considerably complicates the analytical solution of a problem on water resistance of apparatuses with the twirled gas stream. In this connection, now it is necessary to prefer experimental methods of research. In the given work, the problem of definition of water resistance of the irrigated apparatus was put at change of loadings on phases, and also, at a compulsory twisting of a stream, in terms of angular speed of twirl of a rotor and veering of twirl of guide vanes of an air swirler. Research of these factors and creation of a design procedure of water resistance is a momentous problem in decrease in metal consumption of apparatuses and power inputs on gas clearing [11-14].

3.2   EXPERIMENTAL RESEARCH STUDIES OF A DYNAMIC SPRAY SCRUBBER

On the basis of the analysis of builds of modern apparatuses for gas clearing the dynamic spray scrubber build is devised and taken out for a patent. The apparatus is supplied twirled by an air swirler and the central pipe for supply of an irrigating liquid. A centrifugal force originating at twirl of a rotor secures with liquid crushing on microfogs that causes intensive contact of gases and trapped corpuscles to a liquid. Thanks to the act of a centrifugal force, intensive mixing of gas and a liquid, and presence of the big interface of contact, there is an effective clearing of gas in a bubble column [1]

3.2.1 TECHNIQUE OF CONDUCTION OF EXPERIMENT

Research studies of water resistance of a scrubber with an air swirler in altitude 0.25 m were spent on the experimental installation represented in Figure 3.1. Magnitude of hydraulic losses was defined on the difference of static pressure of a gas stream before and after the presence of rotor.

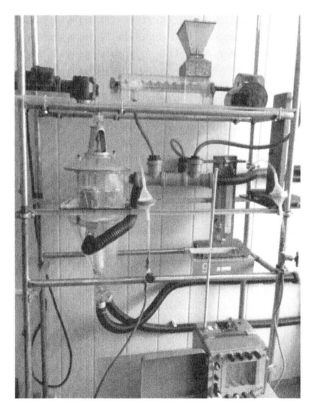

FIGURE 3.1 Experimental installation.

During the course of research the following parameters are varied:

- Speed of the gas on an entry in an air swirler $\upsilon = 1 \div 20$ mps;
- Angular speed of twirl of a rotor $\omega = 0 \div 100$ c^{-1};
- Direction of rotation $\alpha < 90°$ or $\alpha > 90$, where α is an angle between a vector of relative speed ω and a peripheral velocity vector υ;
- Angle of installation of blades $\alpha = 0 \div 65°$

In the first series of experiences, tests by definition dependences of resistance of the apparatus on speed of gas in the apparatus in the absence of liquid supply have been conducted. The air consumption was chosen such to secure with a velocity band, counted for cross-section of the contact channel from 0 to 20 mps. For this purpose by means of the butterfly valve installed, the value of air consumption is measured by means of the gas counter and thus fixed resistance of the apparatus. Measurements were made by means of a pressure gauge. Thus, it has been gained an order of 20 points. The following series of experiences was spent with apparatus irrigation in system water—air. The air consumption was chosen so that to secure with a velocity band, counted for cross section of the contact channel from 5 to 25 mps. The bottom value of a range was defined on the minimum admissible speed of gas which in turn depended on loading on a liquid. A specific irrigation varied within L/G 0–1.5. Each series of experiences was spent at a variable speed of air and a constant irrigation. On the fluxion of 5 min the stationary regime at which following indications were made was installed:

1. Water resistance of the apparatus (Pascal)
2. The charge of a liquid is liter/hour, at conversion on a specific water concentration in the contact channel it is m^3/m^3
3. Gas rate (m^3/h), at conversion for speed of gas in the contact channel (mps).

Pressure indications were made in a gas pipe to the apparatus. For definition of dependence of water resistance from speed of gas in the contact channel at parameter point of a specific irrigation, experiments were spent in a following order.Originally installed value of a specific irrigation, equal to $L/G = 0.2$. Value of loading on a liquid measured by the graduated cylinder. A gas rate of speed in the contact channel changed by means of the butterfly valve. Air consumption is measured by the gas counter. Installed such value of an air consumption at which the liquid undershooting stopped. Thus, the pressure gauge measured an apparatus water resistance. Magnitude of water resistance and the value of a gas rate converted for speed of gas in the contact channel, registered in the table at parameter point of a specific irrigation. Further at gas rate increase the water resistance was again defined.

Analogous experiences have been spent at a parameter point of a specific irrigation equal $L/G = 0.3 - 1.5$. Thus, it has gained an order of 20 points for each of magnitudes of a specific irrigation. The following series of experiences was spent for the purpose of studying the dependence of water resistance from loading a liquid converted in a specific water concentration at constant value of speed of gas in the contact channel. For this purpose value of a gas rate, 50 m³/h was originally installed (conversion for speed of gas in the contact channel has made 5.5 mps) by means of a diaphragm and it was measured by means of the gas counter. At first a pressure gauge resistance indications in dry alternative of work of the apparatus became. And then the water resistance was measured at the installed value of speed of gas for various magnitudes of a specific irrigation of a liquid. Thus, was, it is gained five points for one speed of gas. Similarly, dependence for gas rates 80.5 and 100.4 m³/h has been investigated, at conversion for speed of gas in the contact channel, it has made 12.8 and 18.4 mps accordingly.

3.2.2   ANALYSIS OF RESULTS OF THE EXPERIMENT

Results of research of the irrigated apparatus presented in Figures 3.2 and 3.3 have shown that dependence $\zeta = /(L/G)$ has increasing character and in the given range has univocal association, except for small area at speeds of gas to 5–7 mps.

The analysis of the gained results has allowed to install that growth of magnitude of a specific irrigation in the apparatus at the expense of increase in altitude of a torch of a sprayed liquid does not render considerable effect on a water resistance. This results from the fact that essential impact makes on head losses speed of a gas stream and extent of overlapping by a liquid of a contact zone of the apparatus, thus generated in a zone of an air swirler the formation of a stream does not undergo change at its current along angular coordinate. Liquid-phase effect on a scrubber water resistance was investigated at speed of gas in the twisting device 5–20 mps, relations of mass flow rates of water and gas $L/G = 0.1 - 1.5$. The altitude of a spray of a liquid is equal to 0.15 m.

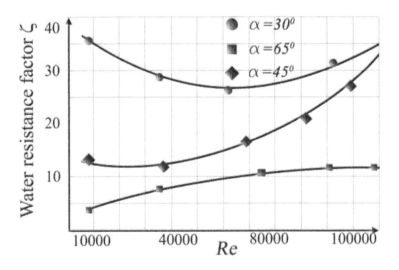

FIGURE 3.2 Dependence of factor of a water resistance on speed of gas and angle of installation of blades of an air swirler.

The essential contribution to the total head losses originating at increase of altitude of a spray of the irrigating liquid is brought in by the losses connected with expenses of energy of a gas stream on transport of a liquid in a zone of twirled guide vanes. Thus, with growth of area of an irrigation these losses will increase. The analysis of results of the experiment, presented in a Figure 3.3, has shown that the water resistance of the irrigated apparatus at low relative loadings L/G drops in comparison with a water resistance of the dry apparatus. Water resistance decrease determines the decrease of tangential making the speed of gas. At the maximum relative loadings, L/G water resistance growth that is connected with expenses originating in the course of work for liquid-phase transport is observed.

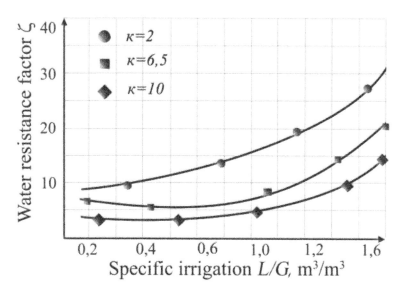

FIGURE 3.3 Dependence of factor of a water resistance on magnitude of a specific irrigation.

During the experiments, it is installed that at a stopping delivery of an irrigating liquid, the water resistance gets not at once value of the dry apparatus, and adopts a value $0.8\Delta P$, and only at disposal of water of a mid-wall of the apparatus and air swirler guide vanes, the water resistance accepts values of the dry apparatus. Water resistance decrease can be proved salutary, affecting the drop layer on the surface of guide vanes of an air swirler, friction decrease about apparatus walls, and as suppression of turbulent pulsations of a gas stream by drops of liquid (Varkasin, 2003).

3.3 WORKING OUT OF A DESIGN PROCEDURE OF A WATER RESISTANCE

3.3.1 EFFECT OF LIQUID PHASE

The water resistance in terms of liquid-phase effects can be expressed as pressure differential sum. Such differences will develop as the resistance originates at motion of gas in the dry apparatus and pressure, which is necessary for informing a gas stream to compensate resistance on transport of a fluid flow stream, d.h.:

$$\Delta P = \Delta P_{dry} + \Delta P_{ir} \qquad (3.1)$$

Or on equation Darcy:

$$\Delta P = \Sigma \xi \cdot \frac{\rho \upsilon^2}{2}$$

$$\Sigma \xi = \xi_{dry} + \xi_{ir}$$

ζ_{dry} is the factor of resistance of not irrigated apparatus; ζ_{ir} is the coefficient of resistance, in terms of the changes which are brought in by an irrigation.

Attempt of theoretical definition of dependence of a head loss of not irrigated apparatus from speed of air by a known technique (Schwidki, 2002) with definition of the sum of factors of local resistance of constructive elements of the apparatus has been made. In attention the following have been accepted: head losses on an entry at turn on 90°, losses on length in terms of stream motions on spirals, and an exit loss from the apparatus. However, comparison of values of the water resistance gained by the above-stated method is considerable above experimental values. It has induced to spend more wide-ranging studies of a water resistance of the "dry" apparatus (Usmanova, 2013). It is installed that the water resistance of the "dry" apparatus submits to quadric dependence on speed of gas. With increase in factor of a twisting ξ decreases that is connected with decrease of level of tangential, making the speed of gas in an air swirler. At feeding into of a liquid phase for some value K, the water resistance factor practically does not depend on the charge of an irrigating liquid that speaks affecting of two factors connected with supply of the irrigating liquid in a dynamic spray scrubber. From 1 party—the increase ξ is connected with growth of head losses of a gas stream on liquid transport; From 2 parties ξ decreases because of decrease in tangential speed.

Gas at the expense of braking act of a liquid

Generalization of experimental data has allowed to gain the equations for water resistance definition in an aspect:

$$\xi_{dry} = \frac{1}{n}\left(\left(\frac{R}{r}\right)^{2n} - 1\right) + \frac{1}{k^2}\cdot\left(\frac{\vartheta_2}{\vartheta_1}\right)^2, \tag{3.2}$$

$$\xi_{ir} = 4\cdot\left(\frac{L}{G}\right)^{0.6}\cdot\sqrt{1+\frac{1}{k^2}}, \tag{3.3}$$

where R is the apparatus radius, m; r is the whirlwind radius, m; L, G are liquid and gas volume flow rates, m³/h; v_1, v_2 are the speeds of gas on an entry and an exit from the apparatus, mps; ε is the factor of loss of a twisting of a stream; K is the factor of a twisting of an air swirler; n is the whirling motion parameter.

The gained expressions for definition of hydraulic losses of the dry apparatus and losses on liquid-phase transport allow to design water resistance in an investigated range of loadings on phases.

3.3.2 EFFECT OF TWIRL OF A ROTOR

At rotor twirl, the investigated vortex generator represents not that other, as the centrifugal fan vane wheel rotor. According to the theory of centrifugal fans (Kostochkin, 1951) magnitude of the theoretical pressure created by twirled rotor, taking into account infinite great number of shovels, is defined by the formula given as follows:

$$H_T\infty = \rho u(u - \frac{v_r}{tg\beta_2}) \tag{3.4}$$

where v_r is the radial component of absolute speed of gas on an exit from a rotor, mps; and β is an outlet angle of a gas stream from a rotor.

A triangle of velocities on an exit from a rotor is shown in Figure 3.4

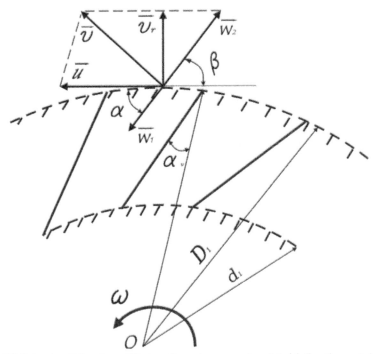

FIGURE 3.4   A roll forming of blades of a vortex generator. A twirl direction: $\alpha \geq 0$.

$$v_r = W_2 \sin \beta_2 \tag{3.5}$$

where W_2 is the relative speed of gas on an exit, mps. The formula of theoretical pressure Eq. (3.1) taking into account and Eq. (3.2) will register as

$$H_T \infty = \rho u(u - W_2 \cos \beta_2) \tag{3.6}$$

However, as appears from special experimental research studies of vane wheel rotors of centrifugal fans (Kostochkin, 1951), theoretical pressure by formula (3.3) reliably enough it is possible to count only for wheels with the big density of a lattice ($\tau_p > 3$). Density of a lattice of wheels of centrifugal fans is defined by the formula

$$\alpha_y > 0 \quad \tau_p = \frac{2l_1 z}{\pi D_1 + l_1 z \sin \alpha_y} \tag{3.7}$$

$$\alpha_y = 0 \quad \tau_p = \frac{l_1 z}{\pi(D_1 + 2l_1)} \tag{3.8}$$

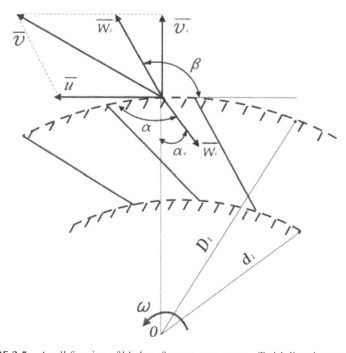

FIGURE 3.5   A roll forming of blades of a vortex generator. Twirl direction: $\alpha \le 0$.

Also is, as a rule, within $\tau_p = 0.8 - 2.5$. Therefore, during the calculation of theoretical pressure correction factor μ is inducted, considering the final number of shovels. For wheels with angles $\beta_2 = 20 + 70°$ factor μ, it is possible to count by formula.

$$\mu = \frac{1}{1 + \dfrac{1,5 + 1,1\beta_2 / 90}{z(1 - \bar{d_1})}} \tag{3.9}$$

Considering Eq. (3.6), formula (3.3) will become

$$H_T = \rho u(u - W_2 \cos \beta_2)\cfrac{1}{1+\cfrac{1,5+1,1\beta_2/90^\circ}{z(1-\bar{d}_1^2)}} \qquad (3.10)$$

At gas motion in a twirled rotor from the center to periphery the theoretical pressure created by a rotor is computed by formula (3.7). The gas stream in a dynamic scrubber vortex generator moves from periphery to its center. Therefore, the gas stream should overcome except a hydraulic resistance of motionless feeding into, also the pressure created by twirled vortex generator in the assumption of a return current of gas through him. At such assumption, considering following relationships (Figure 3.5), it is possible to use formula (3.7) in the following view:

$$u = \omega\frac{D_1}{2} \quad \beta_2 = \alpha$$

$$\Delta\rho_{\omega>0} = \cfrac{0,5\rho\omega D_1(0,5\omega D_1 - W_1 \cos\alpha)}{1+\cfrac{1,5+1,1\alpha/90^\circ}{z(1-d_1^2)}} \qquad (3.11)$$

where d_1 is the relative root diameter of an air swirler, which is designed by either formula (3.8) or Eq. (3.9). Considering Eqs. (3.1) and (3.11), magnitude of water resistance of a dynamic spray scrubber is designed using the semiempirical formula:

$$\Delta P = \Delta P_{dry} + \Delta P_{ir} + \Delta\rho_{\omega>0} \qquad (3.12)$$

3.4   CHECKING ADEQUACY OF THE OFFERED DESIGN PROCEDURE

Results of an experimental research of a hydraulic resistance of a rotor depending on the angular speed of twirl are presented in Figure 3.6. By drawing full lines, experimental values of a hydraulic resistance of a twirled vortex generator, and dotted, which is computed by formula (3.9) are shown. The divergence of desing and experimental values of a hydraulic resistance of a rotor does not exceed 10 per cent.

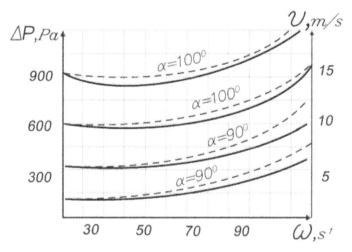

FIGURE 3.6   Dependence of a water resistance on speed and air swirler direction of rotation.

3.5   CONCLUSIONS

1. Experimental and theoretical research studies of hydrodynamics of a scrubber with twirled air swirler are conducted.
2. The research problem made definition of water resistance of the irrigated apparatus at change of loadings on phases, and also, at a compulsory twisting of a stream, in terms of angular speed of twirl of a rotor and veering of twirl of guide vanes of an air swirler.
3. Calculation of a twirled air swirler is offered for designing, being based on the theory of centrifugal fans, in terms of quantities and directions of rotation of guide vanes.
4. The gained expressions for definition of hydraulic losses of the dry apparatus, losses on transport of a liquid phase, and losses at a compulsory twisting of a stream allow to design water resistance in an investigated range of loadings on phases.
5. Research of these factors and creation of a design procedure of water resistance is the momentous problem in decrease in metal consumption of apparatuses and power inputs on gas clearing.

KEYWORDS

- **Guide vanesthe scrubber**
- **The centrifugal fanthe water resistance**
- **The scrubberthe specific irrigation**
- **The specific irrigation**
- **The twirled air swirler**
- **The twirled air swirlerguide vanes**
- **The water resistance the centrifugal fan**

REFERENCES

1. Varkasin, A. J.; Turbulent Flows of Gas with Firm Corpuscles. M.: Engineering; **2003**.
2. Batting, H. A.; Air Purification by Means of Apparatuses of Type the Cyclone Separator. *SPb.* **2003**.
3. Kostochkin, V. N.; Centrifugal fans. Theory and Calculation Bases. M.: Engineering Publishing; **1951**.
4. PATENT NO 2339435 Russian Federations Dynamic Spray Scrubber. Usmanova, R. R.; November 27, **2008**.
5. Stepanov, A. I.; Centrifugal and axial-flow compressors, blowing machines and ventilating fans. The Theory, a Build and Application. M.: Engineering Industry; **1960**.
6. Usmanova, R. R.; Modelling and optimisation of design data of a scrubber. *The Encyclopaedia Eng.-Chem.* **2013**, *6*, 45–50, *(in Russian)*.
7. Schwidki, V. S.; Clearing of gases. The Directory. M.: Heat Power Engineering; **2002**.
8. Shiljev, M. I.; Afonin, N. A.; and Shiljaev, A. M.; Research of process of a dust separation and water resistance in the cascade of direct-flow cyclone separators. Sulfurs. Building. *A Dignit. Tech.* **1999**, *8*, 25–29, *(in Russian)*.
9. Shtokman, E. A.; Air Purification. M.: Publishing House ACB; **1998**.
10. Ekk, B.; Designing and Maintenance Centrifugal and Axial Fans. Translation from German. M.: Technical Publishing; **1959**.
11. Coufont, C.; Forces on spherical particles in terms of upstream flow characteristics. *French Chem. Eng. Cong.* Saint–Nazaire; **2003**, *9*, 1206–1211.
12. Hagesaether, L.; Jacobsen, H. A.; and Svendsen, H. F.; Theoretical analysis of fluid particle collisions in turbulent flow. *Chem. Eng. Sci.* **1999**, *21*, 4749–4755.
13. Hinze, J. O.; Turbulence. New York: McGraw-Hill; **1975**.
14. Schubert, H.; The role of turbulence in unit of particle technology. *World Congr. Part. Technol. (Kuyoto)-Tokyo.* Part 3. **1990**, 55–67.

CHAPTER 4

OPTIMIZATION OF SPEED AND DIRECTION OF TWIRL OF BLADES OF A VORTEX GENERATOR

CONTENTS

4.1   INTRODUCTION

The principle of act of a dynamic spray scrubber is analogous to a principle of act of cyclone separators. In either case allocation of dust from a cleared dust-laden gas stream occurs under the influence of centrifugal force originating at twirl of a stream in the body. Distinctive features of a dynamic spray scrubber consist that main stream curling is carried out by means of tangential feeding into [1–10]. Besides, already twirled gas is exposed to additional curling by means of a twirled rotor.

Advantages of dynamic spray scrubbers in comparison with cyclone separators are higher efficiency of trapping of a finely divided dust, a smaller abrasive wear of internal surfaces of the apparatus, possibility of clearing of gases of higher temperature owing to irrigation by water, and also possibility of regulating of process of separation by change of speed of twirl of a rotor. To defects of dynamic spray scrubbers, it is possible to refer to presence of the additional drive for rotor twirl, the big complexity of the apparatus in manufacturing and maintenance. Owing to complexity of the processes proceeding in dynamic spray scrubbers, the method of their engineering calculation is not devised yet [11–13].

4.2   RESEARCH OF EFFECT OF A DIRECTION OF ROTATION OF AN AIR SWIRLER FOR OPTIMUM SPEED

During the analysis of results of preliminary research studies of a dynamic spray scrubber, it has been noted that with increase in angular speed of twirl of an air swirler, magnitude of ablation of dispersion particles decreases and at certain speed of twirl becomes almost equal to null. Therefore, it is possible to introduce the concept of "optimum speed of twirl," that is, the minimum angular speed of twirl at which there is no reentrainment. It is convenient to operate with optimum angular speed of twirl because it gives the chance to compare skilled values and desing [14–18].

Results of experiences on research of conditions of a twisting and air swirler twirl have shown that optimum speed of twirl a ω_{opt} influences various regime and design data. At studying of effect of an angle of installation of blades of an air swirler on efficiency of separation of disperse streams, it has been installed that for the same angle α, optimum speed of twirl the ω_{opt} depends on the direction of rotation, that is, from an angle $\acute{\alpha}$. As a positive direction of rotation, we will consider a heading, for which cos ά>0. Dependence of ω_{opt} on a twirl cosine of the angle is presented in the form of the schedule (see Figure 4.1).

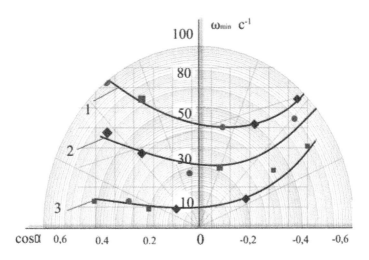

FIGURE 4.1   Dependence of optimum speed of twirl a ω_{opt} from the direction of rotation ά at: (1)Y=21.2; (2) v=13.5; and (3) v=7.7 (mps).

It is visible that for fasted ά, for example, ά=10°, magnitude the ω_{opt} forά>90° is less than for ά<90°. Dependence of ω_{opt}=f(cos ά)has an extreme which is necessary on a negative cosine of the angle of twirl cos ά= −0.2 and can be presented in the following equation ω_{opt}=exp (a cos ά+b cos²ά)

where a, b are factors, which are defined after dependence representation in new coordinates (lg aω_{opt} − lg $\omega_{-0.6}$)/cos α+0.6andcos α. Thus, magnitude $lg\ \omega_{-0.6}$ characterizes the minimum speed of twirl at cos α= −0.6.

After data handling, presented in the schedule (see Figure 4.2), we will gain dependence of optimum speed of twirl on an air swirler direction of rotation.

$$\omega_{opt} \approx \exp\left[-(1,06+0,034v)\cdot\cos\alpha - 2,18\cdot\cos^2\alpha\right] \qquad (4.1)$$

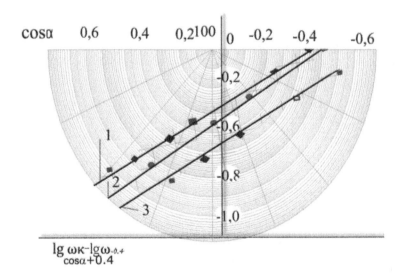

FIGURE 4.2 Effect of cos α on ω_{opt} at a speed of the stream: (1)$v=21.2$; (2)$v=13.5$; and (3)$v=7.7$ (mps).

Thus, by generalizing the results of research of effect of an angle of installation of blades of an air swirler and the direction of rotation, it is possible to draw a conclusion that peak efficiency of separation is secured at installation of blades on an angle $\alpha=10\div30°$ and a negative direction of rotation, that is, $\acute{\alpha}=100+110°$. The question of an aerodynamic roll forming of guide vanes of an air swirler with an estimation of its effect on hydrodynamics and efficiency of process of separation is in detail observed in-process (Usmanova, 2013). It is necessary to note that efficiency of separation raises at an inclination of blades toward air swirler twirl that it is possible to explain decrease of tangential speeds of gas and firm corpuscles; therefore, an inertial force increases. However, the power demand thus increases by the drive that can cause decrease in efficiency of the scrubber.

4.2.1 EXPERIMENTAL RESEARCH STUDIES

On the basis of the analysis of builds of modern apparatuses for gas clearing, the dynamic spray scrubber build is devised and proprietary. The apparatus is supplied by twirled air swirler and the central pipe for supply of an irrigating liquid. A centrifugal force originating at twirl of a rotor secures with liquid crushing on microfogs that causes intensive contact of gases and trapped corpuscles to a liquid. Thanks to the act of centrifugal force, intensive mixing of gas and liquid and presence of the big interface of contact, there is an effective clearing of gas in the bubble column (see Figure 4.3).

FIGURE 4.3 Experimental installation.

All indications were made according to the technique stated in detail in Chapter 3. At research, operating are conditions changed: speed of a gas stream and quantity of the liquid fed on an irrigation of the apparatus. In the capacity of design data, values of number of blades and air swirler critical bucklings are varied.

4.3 RESEARCH OF EFFECT REGIME-DESIGN DATA FOR OPTIMUM SPEED

At discussion of results of preliminary research studies of a dynamic spray scrubber, it has been noted that with increase in angular speed of twirl of a rotor magnitude of ablation of impurity decreases and at certain speed of twirl becomes almost equal to null. Therefore, it is possible to introduce the concept "optimum speed of twirl," that is, the minimum angular speed of twirl at which there is no ablation. Optimum angular speed of twirl is more convenient for using because it gives the chance to compare skilled values and desing.

The number of blades z, presented as a relative of blades $\check{z}=\pi/z$ and equal to a relative step of blades $\tau=(\pi D/z)/D$ essentially influences the efficiency of separation. With increase in number of blades, the minimum speed of twirl decreases (centimeter; Figure 4.4).

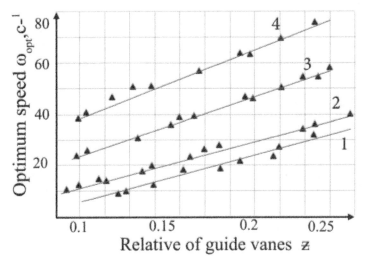

FIGURE 4.4   Effect of a relative of guide vanes on a ω_{opt} at: (1)v=7.2; (2)Y=13.5; (3) v=21.7; and (4)v=34.5 (mps).

It speaks the increase in number of separating channels and distance decrease between channel walls. The path of motion of a corpuscle to walls of the channel decreases and the probability of sedimentation of a corpuscle in the channel raises. However, the number of blades is restricted to conditions of adaptability to manufacture of manufacturing of an air swirler and its working capacity. Proceeding from the spent research studies and in view of conditions of adaptability to manufacture an air swirler, that is number of blades, it is recommended to choose the following relationship:

$$z \approx (100 \div 110)D \qquad (4.2)$$

Generalization of empirical data installs the following relationship:

$$\omega_{opt} \approx \check{z}^{.05} \qquad (4.3)$$

At abbreviation of quantity of blades, activity of affecting the rotor decreases for a gas stream owing to what decrease pressure of gas on an exit and expenses of energy for a transmission of rotary table. From the resulted analysis follows that the increase in number of blades can be expedient including if the increase in pressure of gas at an exit from the apparatus is required and is not a rational way of raise of efficiency of clearing of gas.

Results of experiences on research of a twirled air swirler have shown that optimum speed of twirl, a ω_{opt} influences various gas—liquid and design data. In particular, it is installed for the effect of dispersity of a solid phase at various speeds of a gas stream, see Figure 4.5.

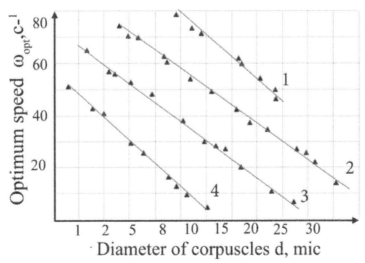

FIGURE 4.5 Effect of diameter of dispersion particles on a ω_{opt} at: (1) υ=34.5; (2) υ=21.7; (3) υ=13.5; and (4) υ=7.2 (mps).

Statistical generalization of empirical data installs a following relationship:

$$\omega_{opt} \approx \exp(-0.018 \cdot 10^6 d_p) \qquad (4.4)$$

Also effect of charges of liquid and gas phases, Figures 4.6 and 4.7 is experimentally installed.

The increase in charge G of the cleared gas at invariable value of angular speed leads to proportional raise of an axial velocity of gas and a decline of separation of a dust owing to decrease of duration of stay of firm corpuscles in air swirler shovels. Considering also that hydraulic losses increase proportionally to a square of an axial velocity of gas, it is expedient to secure with moderate values $\dot{w}$ (to 40 mps).

The increase in optimum speed, a ω_{opt} leads to proportional growth of a gas rate and, hence, to decrease of duration of stay of corpuscles of dust in rotor channels. However, thus it is proportional to ω_{opt}^2 and centrifugal force grows. As a result, separation of firm corpuscles improves. Thus, the increase in angular speed of a rotor raises the efficiency of clearing of gas. However, thus increases the power demand by the apparatus drive (approximately proportionally ω^2) and mechanical loadings on a rotor.

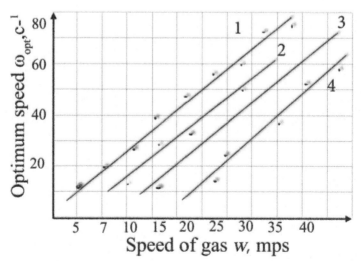

FIGURE 4.6   Effect of speed of a gas stream on a ω_{opt} at: (1) d_p=65; (2) d_p=40; (3) d_p=30; and (4) d_p=10 (μ).

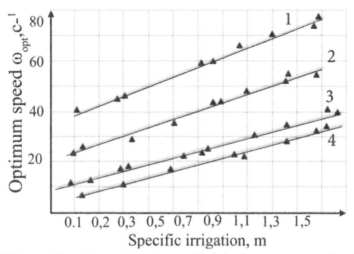

FIGURE 4.7 Effect of magnitude of a specific irrigation on a ω_{opt} at: (1)v=34.5; (2) v=21.7; (3)v=13.5; and (4)v=7.2 (mps).

Statistical generalization of empirical data installs the following relationships:

$$\omega_{opt} \approx w^{1.65} \tag{4.5}$$

$$\omega_{i\,pt} \approx (m \cdot 10^6)^{0.31} \tag{4.6}$$

Except operating conditions, optimum speed of twirl a ω_{opt} influences as well design data of the investigated apparatus. In Figure 4.8, dependence of a ω_{opt} on diameter of a twirled air swirler is illustrated.

The increase in diameter of an air swirler $\check{D}$ on the one hand complicates gas clearing, but with another leads to decrease of an axial velocity of gas and to increase in a time of a finding of firm corpuscles in rotor channels. As a result separation improves. For example, in our research at increase of $\check{D}$ in 1.3 times, the axial velocity has decreased almost 2 times and factor e_0 droppings of a dust has decreased also in 2 times. The rotor volume has essentially increased by 70 percent.

With increase of $\check{D}$ energy loss in a rotor which depends on a gas axial velocity decreases.

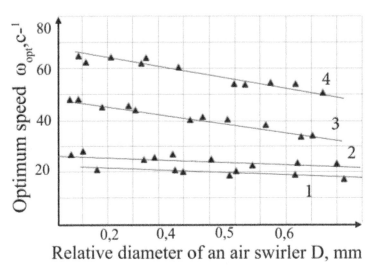

FIGURE 4.8   Effect of design data of an air swirler on a ω_{opt} at: (1) v=7.2; (2) v=13.5; (3) v=21.7; and (4) v=30.7 (mps).

Statistical generalization of empirical data install a following relationship:

$$\omega_{opt} \approx \breve{D}^{-.31} \tag{4.7}$$

The general dependence in terms of Eqs. (4.1–4.7) can be presented in the form of the empirical formula:

$$\omega_{opt} = 397.38 \cdot w^{1.65} (m \cdot 10^6)^{0.31} \breve{D}^{-0.31} \cdot \breve{z}^{1.05}$$
$$\exp\left[-0.018 \cdot 10^6 d_p - (1,06 + 0,034w) \cdot \cos\alpha - 2,18 \cdot \cos^2\alpha\right]$$

The divergence of values of optimum speed of twirl a ω_{opt}, computed to the empirical formula and defined experimentally, does not exceed 13 percent.

4.4 CONCLUSIONS

1. Empirical dependence for calculation of optimum speed of twirl of the air swirler, considering the direction of rotation of its guide vanes, and also operating conditions of process of clearing of gas emissions is gained.

2. The dependences gained by generalization of empirical data give the chance to size up the effect of design data of a dynamic spray scrubber, and also gas and firm corpuscles for optimum speed of twirl a ω_{opt}. Besides, these formulas can be used for preliminary sampling of values of separate design data of a scrubber.

KEYWORDS

- **A swirl angle**
- **Air swirler twirl**
- **Design data**
- **Optimum speed**
- **Quantity of blades**
- **The direction of rotation**

REFERENCES

1. Vatin, N.I.; and Strelez, K.I.; Air purification by means of apparatuses of type the cyclone separator.*SPb*.**2003**.
2. Usmanova, R.R.; Zaikov, G. E.; and Deberdeev, R. I.; Bulletin КТУ.**2013**, *16(7)*, 233–236,*(in Russian)*.
3. Patent, Usmanova, R.R.; The Russian Federation 2339435;**2008**.
4. Varkasin, A. J.; Turbulent flows of gas with firm corpuscles. M.: The Physical-mat; **2(27), 34-49, 2003**.
5. Shvydky, V.S.; Clearing of gases. The directory, M.: Heat Power Engineering; **4(19), 20-31, 2002**.
6. Usmanova, R.R.; and Zaikov, G. E.; Bulletin КТУ.**2012**, *20(15)*, 16–21, *(in Russian)*.
7. Kostockin, V. N.; Centrifugal fans. Theory and calculation bases, M.: Engineering Industry;**1951**.
8. Ekk, B.; Designing and Maintenance Centrifugal and Axial Fans. The lane with it,M.: Technical Publishing House;**1959**.

9. Bitchkov, A. G.; Centrifugal fans for pneumatic transportation of fibrous materials. In: The Collector: Industrial Aerodynamics.M.: Defensive Publishing House;**1957,***9*,91–108 *(In Russian)*.

10. Bitchkov, A. G.; The general regularity of change of aerodynamic characteristics of centrifugal cars with volute casings. In: The Collector: Industrial Aerodynamics.M.: Defensive Publishing House;**1958,***10*, 77–110,*(in Russian)*.

11. Bitchkov, A. G.; Aerodynamic characteristics, areas of work and a drawing for sampling centrifugal and axial fans. In: The Collector: Industrial Aerodynamics.M.: Defensive Publishing House;**1960,***9*, 102–121, *(in Russian)*.

12. Kirillov, B. A.; The Theory of Turbomachines. L: Engineering Industry;**1972.**

13. Kovalenko, V. M.; and Tchebyshev, T. V.; Regulating of centrifugal fans with directing apparatuses on an entry. In: The Collector: Industrial Aerodynamics.M: Defensive Publishing House;**1959,***12,* 70–109,*(in Russian)*.

14. Kovalenko, V. M.; The centrifugal two-level blowing machine of small rapidity. In: The Collector: Industrial Aerodynamics.M: Defensive Publishing House;**1963,** *12,* 108–120, *(in Russian)*.

15. Livshits, S. P.; High-Pressure Cars of Centrifugal Type. L: Engineering Industry;**1976.**

16. Sedov, L. A.; Similarity Methods and Dimensions of a Quantity in the Mechanic. M: The Science;**1972.**

17. Sedov, L. A.; Mechanics of Continuous Medium. T 1 and 2. M: The Science;**1970.**

18. Cebicheva, K. V.; and Solomahova, T. S.; Effect of entrance elements of centrifugal fans on their aerodynamic characteristics. In: The Collector: Industrial Aerodynamics. M.:Engineering Industry;**1974,***31,* 25–39, *(in Russian)*.

CHAPTER 5

EFFECT OF DESIGN DATA OF BLADES OF A VORTEX GENERATOR ON EFFICIENCY OF CLEARING OF GAS

CONTENTS

5.1 INTRODUCTION

Air swirler design data are one of the momentous geometrical sizes of a scrubber and define almost all other sizes. As magnitude of a centrifugal force is inversely proportional to twirl radius, the overall performance depends on the diameter of the scrubber. The diameter of the scrubber, the centrifugal force more developing it and that above its separating ability less [1–8].

Some representation about dependence of productivity of a scrubber and the bottom boundary line of sizes of the corpuscles which are giving into separation, from diameter of an air swirler is given by works (Kouzov, 1993; Prokofichev, 1997; Pirumov, 1961; and Valdberg, 1985). As already it was specified earlier, magnitude of diameter of an air swirler is connected with other geometrical sizes, in particular with length of guide vanes. Magnitude of the last should be not less than twice more than the maximum size of a cone part of an air swirler. It should be meant at definition of diameter of a scrubber.

The increase in altitude of a cylindrical part of a scrubber as well as decrease of an angle of a conic part of an air swirler, increases the dwell(ing) time of corpuscles in a zone of separation and raises the efficiency of clearing of gas emissions [9–12].

5.2 DEVELOPMENT OF MOTION OF A CORPUSCLE IN THE TWIRLED CHANNEL

Now motion of corpuscles in the field of a centrifugal force, characteristic for the twirled stream at tangential gaseous feed, is well enough studied (Idelchik, 1968; Rakhmonov, 2005; Smulsky, 1992; Saburov, 1982; and Saburov, 1989). Therefore, we will observe the motion of corpuscles at the another stage of separation, that is, in shovels of a twirled air swirler. We will consider that the gas-particle system is "the diluted" system [1–8], that is, system, at which distance between corpuscles is much more the diameter of corpuscles ($d_p/lp \leq 1$). In this case as a first approximation, it is possible to consider that the profile of speed of a gas stream is not strained also corpuscles are propelled without collisions with each other. For a single spherical corpuscle, using the general dynamic equation of a point in generalized coordinates (Lagrange's equations), it is possible to write down the following equation:

$$\frac{d}{d\tau}\left(\frac{\partial T}{\partial q_1}\right) - \frac{\partial T}{\partial q} = F \tag{5.1}$$

where T is the corpuscle kinetic energy;

q_1 and q_2 are the generalized coordinates and speeds of a corpuscle;

F_i is the generalized forces;

i is the number of generalized coordinates.

Generalized coordinates or the independent parameters are univocal defining location of a corpuscle, in this case are not that other, as corpuscle coordinates s and n in a curvilinear coordinate system, representing streamlines of gas and orthogonal curves to them. The generalized forces, in observed generalized coordinates s and n, represent forces of resistance Fs and F_n. It was noted earlier that small corpuscles get to air swirler channels. Reynolds number of a corpuscle in this case will be $Re \leq 1$ and the regime of a flow of a corpuscle is defined by the Stokes law [13–17].

Then Eq. (5.1) in a general view will register as system of the equations:

$$\frac{d}{d\tau}\left(\frac{\partial T}{\partial \dot{s}}\right) - \frac{\partial T}{\partial s} = 3\pi\mu d_p \left| W_s - \dot{s} \right|$$

$$\frac{d}{d\tau}\left(\frac{\partial T}{\partial \dot{n}}\right) - \frac{\partial T}{\partial n} = 3\pi\mu d_p \left| W_n - \dot{n} \right| \tag{5.2}$$

where W_s and W_n are projections of speed of gas in the channel on coordinates s and n.

The assumption that gas motion in the twirled channel irrotational motion $W_n = 0$, and speed W_s will be equal to speed of gas in channel W earlier was accepted. Besides, it is necessary to consider that the corpuscle moves in the channel with the speed close by speed of gas and an one-directional with it. Therefore, the system of Eq. (5.2) will assume the following equation:

$$\frac{d}{d\tau}(\frac{\partial T}{\partial \dot{s}}) - \frac{\partial T}{\partial s} = 3\pi\mu d_p |W - \dot{s}|$$

$$\frac{d}{d\tau}(\frac{\partial T}{\partial \dot{n}}) - \frac{\partial T}{\partial n} = 3\pi\mu d_p \dot{n} \qquad (5.3)$$

Let us define corpuscle kinetic energy. For this purpose we will observe corpuscle motion in motionless coordinates x and at and mobile s and n, twirled together with shovels.

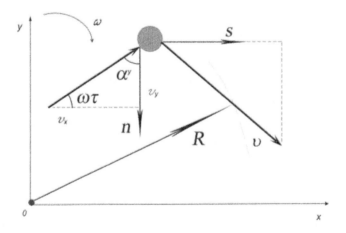

FIGURE 5.1   The loading diagram of location of corpuscles in an absolute motion.

Corpuscle coordinates in an absolute motion (see Figure 5.1), will be the following

$$x = R\cos\omega\tau - S\cos(\omega\tau - \alpha_y) + n\sin(\omega\tau - \alpha_y)$$

$$y = R\sin\omega\tau - S\sin(\omega\tau - \alpha_y) - n\cos(\omega\tau - \alpha_y) \qquad (5.4)$$

Speed of a corpuscle is equal in projections to axes of a motionless coordinate system

$$\dot{x} = -R\omega\sin\omega\tau - \dot{S}\cos\phi + S\omega\sin\phi + \dot{n}\sin\phi + \dot{n}\omega\cos\phi$$

$$\dot{y} = R\omega\cos\omega\tau - \dot{S}\sin\phi - S\omega\cos\phi - \dot{n}\cos\phi + \dot{n}\omega\sin\phi \qquad (5.5)$$

where $\phi = \omega\tau - \alpha_y$.

The corpuscle kinetic energy is defined by the following formula

$$T = \frac{m}{2}(\dot{x}^2 + \dot{y}^2) \qquad (5.6)$$

In terms of Eq. (5.5), the formula (5.6) after transformation will become

$$T = \frac{m}{2} \cdot \begin{bmatrix} R^2\omega^2 + \dot{s}^2 + \dot{n}^2 + \omega \cdot (s + n^2) - 2\omega \cdot (\dot{s}_n - \dot{s}_n) + \\ +2R\omega \cdot (\dot{s}\sin\alpha_y - \dot{n}\cos\alpha_y) - 2R\omega^2 \cdot (s\cos\alpha_y + n\sin\alpha_y) \end{bmatrix} \qquad (5.7)$$

Differentiating (5.7) on coordinates s and n, speeds $\dot{s}$ and $\dot{n}$ and time τ and substituting in system of Eq. (5.2), we will definitively have

$$m(\ddot{s} - s\omega^2 - 2\dot{n}\omega + 2R\omega^2\cos\alpha_y) = 3\pi\mu d_p(W - \dot{s})$$

$$m(\ddot{n} - n\omega^2 - 2\dot{s}\omega + 2R\omega^2\sin\alpha_y) = 3\pi\mu d_p\dot{n} \qquad (5.8)$$

Let us observe corpuscle motion in the channel with flat shovels and we will convert Eq. (5.8)

$$\ddot{s} + A\dot{s} - sw^2 = 2\dot{n}\omega + AW_m + 2A\omega n - 2R\omega^2\cos\alpha_y$$

$$\ddot{n} + A\dot{n} - nw^2 = -2\dot{s}\omega + AW_m - 2R\omega^2\sin\alpha_y$$

where

$$A = \frac{18\mu}{\rho_p d_p^2}; W_m = \frac{Q}{2\pi Rh\cos\alpha_y}$$

also we will lead to its dimensionless form

$$\ddot{\bar{s}} + \dot{\bar{s}} - stk^2\bar{s} + 2stk\dot{\bar{n}} + 2stk\bar{n} - stk\bar{W} - 2stk^2\cos\alpha_y$$

$$\ddot{\overline{n}} + \dot{\overline{n}} - stk^2\overline{n} = 2stk\dot{\overline{s}} - 2stk^2 \sin\alpha_y \qquad (5.9)$$

where

$$\overline{S} = \frac{s}{R}; \overline{n} = \frac{n}{R}; \overline{W} = \frac{W_m}{\omega R}; stk = \frac{\omega d_p{}^2 \rho_p}{18\mu}; \overline{\tau} = \frac{\tau}{\tau_\mu}$$

From the system of differential Eq.(5.9), we will have

$$\ddddot{\overline{n}} + (2stk^2 - 1)\ddot{\overline{n}} + 4stk^2\dot{\overline{n}} + stk^2\overline{n} = 2stk^2 \sin\alpha_y \qquad (5.10)$$

With constant factors, it is possible to present the solution of the non-uniform linear differential equation of the fourth order in the form of the sum of the common decision of Eq. (5.10) without the right part and the private solution with the right part

$$\ddot{\overline{n}} = \dot{\overline{n}}_1 + \dot{\overline{n}}_2 \qquad (5.11)$$

For finding solutions for Eq. (5.10) without the right part we will make eigenvalue equation:

$$k^4 + (2stk^2 - 1)k^2 + 4stk^2k + stk^4 = 0 \qquad (5.12)$$

Equation (5.12) was solved in bundled software Maple 17 within change of criterion of Stokes from $0.4 \cdot 10^{-4}$ to 1.5 As a result it was revealed that at $0.3 \le Stk \le 1.5$, equation has two real roots $()$ and $(-in)$ and two interfaced imaginary $(\alpha \pm \beta)$, then the common decision of differential Eq.(5.10) will have the following appearance:

$$\overline{n} = C_1 \exp(-a\overline{\tau}) + C_2 \exp(-B\overline{\tau}) + (C_3 \cos\beta\overline{\tau} + C_4 \sin\beta\overline{\tau}) \cdot \exp(\alpha\overline{\tau}) \quad (5.13)$$

where C_1, C_2, C_3, C_4 are arbitrary constants; τ is the dimensionless time.

$$\overline{\tau} = \frac{\tau}{\tau_\mu}$$

τ_μ is the corpuscle relaxation time

$$\tau_\mu = \frac{18\mu}{d_p^2 p_p}$$

The partial solution of Eq. (5.10) will register

$$\bar{n}_2 = 2\sin\alpha_y \qquad (5.14)$$

and Eq. (5.11) will register

$$\bar{n} = C_1\exp(-a\bar{\tau}) + C_2\exp(-B\bar{\tau}) + (C_3\cos\beta\bar{\tau} + C_4\sin\beta\bar{\tau})\cdot\exp(\alpha\bar{\tau}) + 2\sin\alpha_y \quad (5.15)$$

Substituting Eq. (5.15) in system of Eq. (5.9) and solving rather $\dot{s}$, we will have

$$
\begin{aligned}
\bar{s} = {} & AC_1\exp(-a\bar{\tau}) + BC_2\exp(-b\bar{\tau}) + \\
& + \left[(CC_3 + DC_4)\cos\beta\bar{\tau} + (CC_4 + DC_3)\sin\beta\bar{\tau}\right]\cdot\exp(\alpha\bar{\tau}) + \\
& + \frac{\bar{W} + 2stk\cos\alpha_y - 4\sin\alpha_y}{Stk}
\end{aligned}
\qquad (5.16)
$$

where

$$A = \frac{-a^3 + 3astk^2 + 3stk^2}{2stk^3}$$

$$B = \frac{-b^3 + 3bstk^2 - 3stk^2}{2stk^3}$$

$$C = \frac{-a^3 + 3\alpha\beta^2 + \alpha - 3\alpha stk^2 - 3stk^2}{2stk^3}$$

$$D = \frac{\beta^3 - 3\alpha^2\beta + \beta - 3\beta stk^2}{2stk^3}$$

Using the starting conditions:

$$\bar{n} = 0 \,; \dot{\bar{n}} = 0 \,; \; \bar{s} = 0 \,; \dot{\bar{s}} = stk\overline{W}$$

at $n=0$, we will make system of four equations for search of arbitrary constants C_1, C_2, C_3, and C_4

$$C_1 + C_2 + C_3 = -2\sin\alpha_y$$

$$aC_1 + bC_2 + \alpha C_3 + \beta C_4 = 0 \qquad\qquad (5.17)$$

$$a^2 C_1 + b^2 C_2 + (\alpha^2 - \beta^2)C_3 + \alpha\beta C_4 = -2stk^2(2stk\cos\alpha_y + \overline{W} - \sin\alpha_y)$$

$$a^3 C_1 + b^3 C_2 + (\alpha^3 - \alpha\beta^3)C_3 + (3\alpha^2\beta - \beta^3)C_4 = 2stk^2(2stk\cos\alpha_y + \overline{W} - \sin\alpha_y)$$

The system of Eq. (5.17) was solved in bundled software Maple 17 with the following changes of magnitudes entering into it:

$0.3 \leq Stk \leq 1.5$; $W=0.4\div5$; $sin\alpha_y=0\div0.8$; $cos\alpha_y=0.6\div1.0$

The found values of arbitrary constants were substituted in Eqs. (5.15) and (5.16).

5.3   DEFINITION OF EFFECTIVE LENGTH OF BLADES

It is possible to consider that the corpuscle in twirled shovels of an air swirler will be separated, if it attains a wall of blades. This condition allows to design effective length of blades h_{eff} in which the corpuscle of the set size is separated (Figure 5.2).

For definition of effective length of blades, it is necessary from Eq. (5.17) at

$$\bar{n} = -\bar{z}\cos\alpha_y$$

$$\bar{z} = \frac{\pi}{z}$$

So, the product of four of them, namely Eqs. (5.22), (5.23), (5.24), and (5.28) gives not that other, as the relation of a Reynolds number of a gas stream in the field of a dispensing of an irrigation water and an irrigation water stream, at an exit from the sprinkler hole, increased by a dimensionless quantity.

Results of calculation h_{eff} for $\hat{z}$=0.05÷0.5 show that in the accepted limits of change of criterion of Stokes, speed of gas in shovels and angular speed of twirl of an air swirler magnitude of effective length of blades are negative. It means that corpuscles at values of criterion of Stokes more than 0.3 are separated in a clearing initial stage. Really, at μ=1.85·10^{-5} a Pascal s, ρ=10^3 kg/m^3 and $Stк$=0.3; sizes of corpuscles are big enough d_p=180·10^{-6} m for ω=10 c^{-1} and d_p=50·10^{-6} m for ω=100 c^{-1}.

From schedules it is visible that with the decrease of criterion of Stokes magnitude of effective length of a shovel sharply increases and at $Stк \leq 0.05$ tends to infinity. Hence, corpuscles for which $Stк \leq 0.05$ practically cannot be trapped in a twirled air swirler.

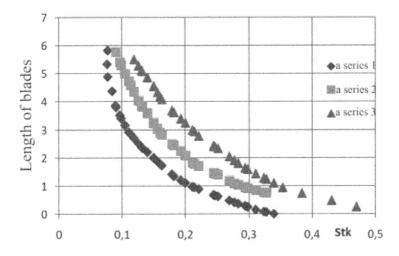

A series 1:α_y= −20°, v=5; A series 3:α_y=0°, v=10; A series 3:α_y=10°,v=15;

FIGURE 5.2   Dependence of length of blades of an air swirler on operating conditions.

All calculations obtained earlier are fair at a direction of rotation of blades α≤90°. Results of calculation for α≥90° show that the effective length of a shovel is slightly less than at a heading α≤90°.

On the basis of the obtained results it is possible to define effective length of twirled blades of the air swirler, necessary for trapping corpuscles of the set diameter at the set speed of gas and angular speed of twirl of an air swirler.

5.4   CONCLUSIONS

1. Motion of corpuscles in shovels of a twirled air swirler in a curvilinear coordinate system, representing streamlines of gas and orthogonal curves to them is observed. The corpuscle kinetic energy is defined.
2. Calculation on the treated equations has shown that in the accepted limits of change of criterion of Stokes, speed of gas in shovels and angular speed of twirl of an air swirler magnitude of effective length of blades are negative.
3. On the basis of the obtained results, it is possible to define the effective length of twirled blades of the air swirler, necessary for trapping of corpuscles of set point diameter at the set speed of gas and angular speed of twirl of an air swirler.

KEYWORDS

- **A particle path**
- **Criterion of stokes**
- **Effective length of blades**
- **Operating conditions**
- **The absolute motion**
- **The kinetic energy**

REFERENCES

1. Belevitskij, A. M.; Designing a Gas-Cleaning Installation of Constructions. L: Chemistry; **3, 23-45, 1990.**
2. Idelchik, I. E.; Alexanders, V. P.; and Kogan, E. I.; Research of direct-flow cyclone separators of system of an ash collection of a state district power station.*Heat Power Eng.* **1968,**8, 45–48,*(in Russian).*

3. Kouzov, P. A.; Malgin, A. D.; and Skryabin, M.; Clearing of gases and air of a dust in the chemical industry. SPb: Chemistry;**1993**.

4. Kutateladze, S. S.; Gyroscopes, E. P.; and Terekhov, V. I.; Aerodynamics in the Restricted Vortex Flows. Novosibirsk: An Academy of Sciences of the USSR;**1987**.

5. Kutateladze, S. S.; and Styrikovich, M. A.; Hydraulics of Gaso-Liquid Systems. State Power Publishing House;**1958**.

6. Levich, V. G.; Physical and Chemical Hydrodynamics. Fizmatgiz;**1959**.

7. Potapov, I. P.; and Kropp, L.D.; Batarejnye Cyclone Separators. M.: Energy;**1977**.

8. Prokofichev, N. N.; Reznik, V. A.; Aleksandrovich, E. I.; and Yermolaev, V. V.; To Sampling of the Ash Collector for Coppers of Industrial and Municipal Power Engineering.**1997**,*8*, 12–13 p, *(in Russian)*.

9. Pirumov, A. I.; Aerodynamic Bases of the Inertia Separation. M: A State Power Publishing House;**1961**.

10. Rakhmonov, T. Z.; Salimov, Z. S.; and Umirov, R. R.; Wet Clearing of Gases in Apparatuses with a Mobile Nozzle. T.: The Fan;**2005**.

11. Straus, V.; Industrial Clearing of Gases. Moscow: Chemistry;**1981**.

12. Sazhin, B. S.; and Gudim, L. I.; Dedusters with the Counter Twirled Streams the Chemical Industry. Moscow;**1984**,*8*, 50–54 p, *(in Russian)*.

13. Smulsky, I. I.; Aerodynamics and Processes in Whirlpool Chambers. Novosibirsk: "Science";**1992**.

14. Saburov, E. N.; Aerodynamics and Convective Heat Transfer in Cyclonic Heating Installations. Leningrad: Publishing House;**1982**.

15. Saburov, E. N.; Carps, S. V.; and Ostashev, S. I.; Heat Exchange and Aerodynamics of the Twirled Stream in Cyclonic Devices. Leningrad: Publishing House;**1989**.

16. The Directory after a Heat and to an Ash Collection. Eds. Birger, M. I.; Valdberg, A. J.; Mjagkov, B. I.; et al. Under Rusanova's, A. A.; general edition. Atom Publishing House;**1983**.

17. Valdberg, A. J.; Isjanov, L. M.; and Tarat, E. J.; Dust Separation Production Engineering. L.: Engineering Industry;**1985**.

CHAPTER 6

SAMPLING OF AN OPTIMUM RULE OF AN IRRIGATION CANAL FOR LIQUID SUPPLY IN THE APPARATUS

CONTENTS

6.1   INTRODUCTION

Spraying ("crushing") of a liquid is widely applied in modern technics. It is carried out, in particular, in chemical and the food-processing industry at extraction of firm substances from liquids, at drying, at any interactings between liquids and gases, and also in a number of other processes (mash crushing in the aluminium industry, cooling of gases by the sprayed liquid in a number of apparatuses etc.) (Vereschagin, 1959; Vitman, 1953; Idelchik, 1968; and Kutateladze, 1958). So widespread application of spraying speaks that in all these processes decrease of sizes of drops sharply increases a surface-area factor and, hence, reduces a time of a leakage of process that allows to reduce gabarits of apparatuses considerably [1–12]. Besides, spraying secures with the big uniformity of distribution of a liquid and its best interacting with the reacting medium.

Numerous experimental researches show that the streams of a liquid outflowing from a hole on medium of gas, pulse (Vitman, 1953; Vitman, 1961; and Miesse, 1955). Under certain conditions the liquid pulsation fade-ins along a stream and leads to its disintegration on drops [13–19]. Character of intermittent motion depends on the nozzle form from which the stream, liquid index of turbulence in flush, physical properties of a liquid and gas and their relative speed outflows [20–28].

Problem about disintegration of streams by means of consideration of stability of the given current of a liquid. Mathematical research of stability of motion in relation to perturbations can be solved by means of the motion equations (Vitman, 1961; and Pirumov, 1961). With that end in view on stationary main current the non-stationary perturbation is imposed so that resulting motion fulfilled to the motion equations. At an outflow velocity having practical interest, gravity effects on liquid motion can to be considered. In this case on a liquid stream forces of viscosity, a superficial tension and a seepage force act [29–35].

6.2   DEVELOPMENT OF MOTION SPHERICAL DROPS OF LIQUID IN A GAS STREAM

At optimum location of a sprinkler the best crushing of a liquid and overlapping by a spray of an irrigation water of all cross-section of the apparatus will be secured.

Let us observe motion of drops of a liquid in the twirled gas stream at optimum location of a sprinkler Θ_0. For simplification of the solution of a problem we will accept following assumptions: we will consider that the drop is in the form of a sphere and is absolutely a solid. Actually the drop form is not strictly spherical and in the course of motion in it originate the deformation caused by force of resistance of gas and counteraction of this deformation, called by a liquid surface tension force (Turner, 1953; and Probert, 1946).

Motion spherical drops of liquid in a gas stream is presented by the vector differential equation in an aspect:

$$m_0 \frac{dV}{dt} = -\psi_c \pi (\frac{D_0}{2})^2 \frac{\rho_r}{2} |U| U + m_0 G, \qquad (6.1)$$

where m_0, D_0—weight and diameter of a drop;
ρ_g—gas density;
V, U—vectors of a relative and absolute traverse speed of a drop;
$/U/$—the module of a vector of relative speed;
G—a vector of acceleration of a gravity;
ψ_c—coefficient of resistance to corpuscle motion, function

$$\mathrm{Re}_v = \frac{|U| D_0}{v_r}$$

It is defined or on an experimental curve (Kutateladze, 1958), or under empirical formulas (Pirumov, 1961);
v_g—kinematic viscosity of gas.
Having increased both sides of an Eq. (6.1) on

$$\frac{6}{\pi \cdot v_r^2 \rho_r}$$

Let us gain

$$\frac{\rho_l D_0}{\rho_r v_r^2} \cdot \frac{dV}{dt} = -\frac{3}{4}(\frac{D_0}{2})^2 \psi_c |U| U + \frac{\rho_l D_0^3}{\rho_r v_r^2} G \qquad (6.2)$$

index

$$\dot{V} = \dot{v}_x i + \dot{v}_y i = \frac{VD_0}{V_r} = \frac{v_x D_0}{V_r} i + \frac{v_y D_0}{V_r} i \tag{6.3}$$

$$\dot{W} = \dot{W}_x i + \dot{W}_y i = \frac{WD_0}{V_r} = \frac{W_x D_0}{V_r} i + \frac{W_y D_0}{V_r} i \tag{6.4}$$

$$\dot{W} = \dot{W}_x i + \dot{W}_y i = \frac{WD_0}{V_r} = \frac{W_x D_0}{V_r} i + \frac{W_y D_0}{V_r} i \tag{6.5}$$

Let us spread out the Eq. (6.2) on components on X-axes and y

$$\frac{\rho_l D_0}{\rho_r v_r^2} \cdot \frac{dv_x}{dt} = -\frac{3}{4}(\frac{D_0}{V_r})^2 \psi_c |U| U_x \tag{6.6}$$

$$\frac{\rho_l D_0}{\rho_r v_r^2} \cdot \frac{dv_y}{dt} = -\frac{3}{4}(\frac{D_0}{V_r})^2 \psi_c |U| U_y + \frac{\rho_l D_0^3}{\rho_r v_r^2} \cdot g \tag{6.7}$$

Let us restructure the left parts of the Eqs. (6.6) and (6.7) for what we will increase and we will divide a derivative in the first equation on dx, in the another on dy, then:

$$\frac{\rho_l D_0^3}{\rho_r v_r^2} \cdot v_x \cdot \frac{dv_x}{d\tilde{o}} = -\frac{3}{4} \cdot \frac{D_0^2}{V_r^2} \psi_c |U| U_x \tag{6.8}$$

$$\frac{\rho_l D_0^3}{\rho_r v_r^2} \cdot v_y \cdot \frac{dv_y}{d\tilde{o}} = -\frac{3}{4} \frac{D_0^2}{V_r^2} \psi_c |U| U_y + \frac{\rho_l D_0^3}{\rho_r v_r^2} \cdot g \tag{6.9}$$

index

$$T = \frac{\rho_r v_r t}{\rho_l D_0^2}; X = \frac{\rho_r x}{\rho_l D_0}; Y = \frac{\rho_r y}{\rho_l D_0};$$

$$P = \frac{\rho_l D_0^3}{\rho_r v_r^2} g; \psi = \frac{3}{4}\psi_c |U|$$

And in view of earlier accepted designations, from the Eqs. (6.6) and (6.7) we will gain the dimensionless equations:

$$\frac{d\dot{v}_x}{dT} = -\psi \dot{U}_x \qquad (6.10)$$

$$\frac{d\dot{v}_y}{dT} = -\psi \dot{U}_y + P \qquad (6.11)$$

and from Eqs. (6.8) and (6.9) accordingly

$$\dot{U}_x \frac{d\dot{v}_x}{dx} = -\psi \dot{U}_x \qquad (6.12)$$

$$\dot{U}_y \frac{d\dot{v}_y}{dy} = -\psi \dot{U}_y + P \qquad (6.13)$$

By means of the gained Eqs. (6.10–6.13) it is possible to define speed of a drop at any moment and to design a path of its motion in a gas stream, moving with a variable speed if these equations to present in final differences.

So, from Eqs. (6.10) and (6.11) it is had:

$$\Delta \dot{U}_y = \Delta \dot{U}_x \left(\frac{\dot{U}_y}{\dot{U}_x} - \frac{P}{\Psi \dot{U}_x} \right), \qquad (6.14)$$

and from Eqs. (6.12) and (6.13) accordingly

$$\Delta X = -\frac{\dot{v}_x \Delta \dot{v}_x}{\Psi \dot{v}_x} \qquad (6.15)$$

and

$$\Delta Y = -\frac{\dot{v}_y \Delta \dot{v}_y}{\Psi \dot{v}_y - P} \qquad (6.15a)$$

The analysis of dimensionless quantities Eqs. (6.10–6.13) shows that compared motions of drops in geometrically similar sprinklers will be similar, if

$$\frac{\rho_r l}{\rho_l D_0} = idem,$$

(6.16)

$$\frac{UD_o}{V_r} = idem,$$

(6.17)

$$\frac{UD_o}{V_r} = idem,$$

(6.18)

$$\frac{\dot{U}_x}{\dot{U}_y} = idem,$$

here l—length of an irrigation pipe.

In this case mechanical trajectories of volumes of a liquid also will be similar (Sedov, 1954).

The time of motion of a drop is subject to the condition:

$$T = \frac{\rho_r V_r t_l}{\rho_l D_0^2}$$

(6.19)

Applying the gained expressions to observed motion of a stream of an irrigation water in a sprinkler and neglecting stream weight, we will gain as a first approximation:

$$\frac{\rho_r d_x}{\rho_l D_0} = idem,$$

(6.20)

$$\frac{U d_l}{V_r} = idem,$$

(6.21)

$$\frac{\dot{U}_x}{\dot{U}_y} = \frac{\omega_{i\,pt}}{v_l} = idem,$$

(6.22)

here d_x—diameter of the conic dissector in the field of a sprinkler bevel, m;

d_1—diameter of a hole of a sprinkler, m;

ω_{opt}—speed of gas in the field of an irrigation, mps;

v_1—a projection of speed of an irrigation water to a perpendicular to a gas traffic route

In this case we accept that all irrigation water was completely converted to flat radial streams.

Thus performance of conditions of similitude Eqs. (6.20–6.22) for compared streams of an irrigation water is necessary, but insufficient. These conditions do not consider liquid supply in the device in the form of the flat radial streams, secured by the conic dissector.

On Figure 6.1 the way of corpuscles of a surface of a cone is represented

In polar co-ordinates R_1 and Θ. By an arrow the flow core direction of rotation is shown. The curve 1 represents cone boundary line, a curve 2—a corpuscle way. In case of the flat dissector the curve 2 represents the valid way of a corpuscle of the liquid arriving on a tangent on a round of a cone and outflowing in its apex.

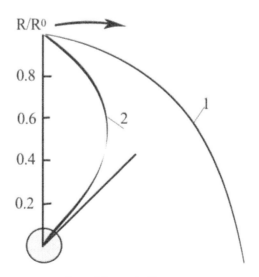

FIGURE 6.1   Drop way on a surface of the conic dissector.

Analysing process of separation of a liquid in a gas stream, the author (Tresch, 1954). Specifies that at drop disintegration acts following forces:

$$\rho_l u^2 D_0^2 ; \sigma D_0 ; \mu_l u D_0 H \rho_r u^2 D_0^2 ;$$

The condition of disintegration of a drop under the influence of a gas stream can be presented in the form of following criteria:

$$\frac{\sigma}{\rho_l u^2 D_0} = idem \qquad (6.23)$$

and

$$\frac{\mu_l}{\rho_r u D_0} = idem \qquad (6.24)$$

Passing round the had dependences to a case of disintegration of continuous streams of a liquid under the influence of a gas stream as a first approximation we will have:

$$\frac{\sigma}{\rho_l u^2 d_l} = idem \qquad (6.25)$$

and

$$\frac{\mu_l}{\rho_r u d_l} = idem \qquad (6.26)$$

Using the had dependences Eqs. (6.25) and (6.26) it is possible to tell that compared motions of streams of the irrigation water which is getting out a sprinkler in geometrically similar apparatuses with axial input of an irrigation water will be similar, if

$$\frac{\rho_r d_x}{\rho_l d_l} = idem \qquad (6.27)$$

$$\frac{U d_l}{V_l} = idem \qquad (6.28)$$

$$\frac{\omega_{opt}}{v_l} = idem \tag{6.29}$$

$$\frac{\sigma}{\rho_l \omega_{opt}^2 d_l} = idem \tag{6.30}$$

$$\frac{\mu_l}{\rho_l \omega_{opt} d_l} = idem \tag{6.31}$$

The analysis of these criteria shows that they are the connected magnitudes since at maintenance of the best crushing of a liquid and full overlapping of cross-section of the apparatus by an irrigation spray, change of one of them involves by all means change and others.

6.3 SAMPLING OF CRITERION OPTIMUM A SPRINKLER RULE

Let us assume that defining criterion Θ_0 will be any other criterion merging had criteria all here.

Therefore, product of four of them, namely Eqs. (6.27), (6.28), (6.29), and (6.31) gives not that other, as the relation of a Reynolds number of a gas stream in the field of a dispensing of an irrigation water and an irrigation water stream, at an exit from the sprinkler hole, increased by a dimensionless quantity, equal

$$\sqrt{1 + (v_l / \omega_{opt})^2}\ ,$$

d.h.

$$\frac{d_x \omega_{opt}}{v_r} \cdot \frac{\mu_l}{\rho_l v_l d_l} \cdot \sqrt{1 + (v_{l\rho_l} / \omega\ )^2} = idem \tag{6.32}$$

For the devised build of a sprinkler for supply of an irrigation water magnitude

$$\sqrt{1+(v_l / \omega_{opt})^2} = f_1(m_1)$$

Therefore expression (6.32) can be presented in an aspect

$$\frac{\text{Re}_r}{\text{Re}_l} f_1(m_1) = idem \qquad (6.33)$$

Here m_1—magnitude of a specific irrigation, equal to the relation of the charge of a liquid to the volume flow rate of gas taken under a condition on an entry in a scrubber, l/m³.

Having denoted the relation of numbers Re_r and Re_l through $\acute{m}$, d.h.

$$\breve{m} = \frac{\text{Re}_r}{\text{Re}_l} = \frac{\omega_{ipt} d_x v_l}{v_l d_l v_r} \qquad (6.34)$$

Expression (6.33) can be written down in an aspect

$$\breve{m} = f_2(\frac{1}{m_1}) \qquad (6.35)$$

were

$$f_2(\frac{1}{m_1}) = \frac{idem}{f_1(m_1)} \qquad (6.36)$$

For the chosen build of a sprinkler for supply of an irrigation water and at use of any one liquid with $\sigma = const$ the criterion Eq. (6.30) will assume the following air:

$$\frac{G_0}{\omega_{opt}^2} = idem, \qquad (6.37)$$

were

$$G_0 = -\frac{\sigma}{\rho_l d_l}$$

Having increased and having divided expression (6.37) on v_1^2, we will have

$$\frac{G_0}{v_l^2} \cdot (\frac{v_l}{\omega_{opt}})^2 = idem \qquad (6.38)$$

The analysis of expression (6.38) shows that the relation $v_l/\omega_{opt} \approx m_1$ (by definition), and $v_l \approx m_1$ at $\omega_{opt} \approx const$ the criterion Eq. (6.30) under the conditions specified above also will be proportional to a specific irrigation.

Research of criteria Eqs. (6.30) and (6.35) shows that as in the capacity of an irrigation water first from them is more the general is usually used and consequently there are all bases to assume that as a first approximation it *(ḿ)* will be defining at a finding of optimum location of feeding into of a liquid through a sprinkler, answering to the best crushing of an irrigation water and more its uniform distribution on apparatus cross-section at preset values m_1 and ω_2. Here ω_2—speed of a gas stream in observed cross-section of a scrubber, mps.

Having defined experimentally dependence of criterion of optimum location of a connecting pipe of supply of an irrigation water *(ḿ)* from a specific irrigation, it will be possible to find by means of this dependence an optimum place of feeding into of a liquid and at others, distinct from experimental, parametres of gas and fluid-flow streams.

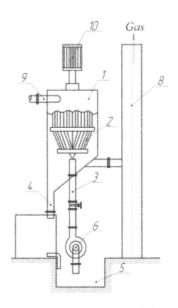

FIGURE 6.2   The circuit design of experimental installation.

So, for the conic dissector observed in-process (Figure 6.2 see) it is had:

$$d_x = d_2 (1 + 2\Theta_0 tg \frac{\alpha_1}{2})$$ (6.39)

and

$$\omega_{i pt} = \frac{4Q_1}{\pi d_2^2 (1 + 2\Theta_0 tg \frac{\alpha_1}{2})^2}$$ (6.40)

where Q_1—a second gas rate at parametres on an entry in the apparatus;
α_1—An angle of disclosing of a cone of the conic dissector.

Substituting the had dependences d_x and ω_{opt} in the formula (6.34) and solving rather Θ_0, we will have

$$\Theta_0 = \frac{L_x}{d_2} = \frac{1}{2tg \frac{\alpha_1}{2}} (\frac{Q_1 v_l d_l}{Q_l v_r \breve{m} d_2 \sin \phi} - 1)$$ (6.41)

where φ—an angle between a heading of outflow of a liquid from a sprinkler and a traffic route of a gas stream.

In that case when Q_1 it is expressed through ω_2, and Q_1 through v_1, the formula (6.41) will assume the following air:

$$\Theta_0 = \frac{1}{2tg \frac{\alpha_1}{2}} (\frac{\omega_2 d_2 v_l \gamma_2}{v_l d_l v_r \gamma_1 \breve{m}} - 1)$$ (6.42)

Where Y_1 and Y_2—relative density of gas at temperature and an inlet pressure and an exit from the apparatus, accordingly, kg/m³.

Optimum location of a sprinkler Θ_0 can be defined also or from expression (6.43)

$$\Theta_1 \frac{L_0}{d_1} = \frac{L_1 - \Psi d_2}{d_1}$$ (6.43)

Or from expression (6.44)

$$\Theta_1 = \frac{1}{2tg\dfrac{\alpha_1}{2}} \left(1 - \frac{Q_1 v_l d_l}{Q_l v_r \breve{m} d_1 \sin \phi} \right)$$ (6.44)

where L_1—total length of an irrigation connecting pipe, m;

d_1—diameter of a sprinkler, m;

L_0—distance from the conic dissector to an irrigation connecting pipe, m;

ψ—the relation of magnitude L_0 to d_1

Research of experimental data (Nukiyama, 1938; Radcliffe, 1953; and Miesse, 1955) about character of a deviation various to diameter and an outflow velocity of streams of water under the influence of a gas stream, moving with various speed, shows that inferred above the formula and criterion scores for a finding of optimum location of a sprinkler Θ_0 for liquid supply can be passed round, apparently, only for the zone of an irrigation restricted on a size. Definition of the maximum zone of the irrigation served by an axial sprinkler, at the moment a theoretical way is not obviously possible and, therefore, it should be made experimentally. Quite probably that for industrial scrubbers at which the apparatus cross-section will be more than the maximum spray of an irrigation formed by one connecting pipe, uniform water delivery on cross-section and provision of full shutdown of a plane by a liquid can be had only at installation of several sprinklers. Definition of optimum location of sprinklers in this case can be defined, apparently, from the expression similar Eq. (6.34), if instead of d_x in the formula (6.34) to substitute d_1

$$\breve{m} = \frac{\omega_{opt} d_i v_l}{v_l d_l v_r} = \frac{\omega_{opt} d_x v_l}{v_l d_l v_r \sqrt{n}}$$ (6.45)

where $d_1 = d_x \sqrt{n}$—diameter i—й the irrigation zones, served by one sprinkler; n—number of installed sprinklers of the central water delivery securing its uniform distribution on cross-section of the apparatus.

Solving the Eq.(6.45) rather Θ_0 and Θ_1, we will have

$$\Theta_0 = \frac{1}{2tg\dfrac{\alpha_1}{2}} \left(\frac{Q_1 v_l d_l}{Q_l v_r \breve{m} d_2 \sqrt{n} \sin \phi} - 1 \right)$$ (6.46)

$$\Theta_0 = \frac{1}{2tg\frac{\alpha_1}{2}}\left(\frac{\omega_2 v_l d_2 \gamma_2}{d_l v_l \breve{m} v_r \gamma_r \sqrt{n}} - 1\right) \tag{6.47}$$

$$\Theta_1 = \frac{1}{2tg\frac{\alpha_1}{2}}\left(1 - \frac{Q_l v_l d_l}{Q_l v_r \breve{m} d_2 \sqrt{n}\sin\phi}\right) \tag{6.48}$$

6.4 CONCLUSIONS

1. By comparison of the forces causing convergence of corpuscles it is shown that capture of trapped corpuscles of a dust by water drops in a scrubber is carried out at the expense of inertia for corpuscles $d \geq 1$ *mic* and diffusions for corpuscles $d \leq 1mic$.

2. At water delivery in the central sprinkler, a place of its feeding into essential impact makes on extent of a dispensing of water, its distribution on apparatus cross-section, creation of a flat spray of an irrigation and efficiency of trapping of a dust. There is an optimum location of a sprinkler Θ_0 at which for set $\acute{m}$ and ω_2 in the apparatus the best crushing of water, more its uniform distribution on cross-section of the apparatus and the highest efficiency of a dust separation is secured.

3. On the basis of the equation of motion of volume of a liquid in gas medium the criterion defining optimum location of a sprinkler is found and experimentally confirmed.

KEYWORDS

- A particle path
- An irrigation spray
- Criterion of defining the optimum
- Fractional coagulation
- Optimum location
- The conic dissector

REFERENCES

1. Blinov, V. I.; and Fejnberg, E. L.; About a Pulsation of a Stream and its Rupture on Drops. GTF.**1933**,*III*, exhaustion. 5.
2. Vasilevsky, M. B.; and Zykov, E.G.; Methods of raise of efficiency of systems of dust removal of gases with group cyclonic apparatuses in small power engineering. *Ind. Power Eng*.**2004**,*9*, 54–57, (in Russian).
3. Weber, K.; Disintegration of a stream of a liquid. The Collector Explosive motors. Ed. Vasileva, S. N.; Works of Academy of Sciences of the USSR;**1936**.
4. Vereschagin, L. Ф; Semerchan, I. A; and Sekojan, S. C.; To a Question on Disintegration of a High-Speed Water Stream. Works of Academy of Sciences of the USSR.**1959**, *XXIX*, 1.
5. Vitman, L. A.; Spraying of a viscous fluid by injectors not centrifugal type. In: The Collector of Scientific Works. Chemistry: Leningrad; **1953**.
6. Vitman, L. A.; About calculation of length of a continuous part at disintegration of a stream of a liquid. In: The Collector Questions of a Convective Heat Exchange and Hydraulics Two-Phase Agent. A State Power Publishing House;**1961**.
7. Gupta, A. D.; and Sajred, H. The Twirled Streams. M.: The World;**1987**.
8. Idelchik, I.E.; Alexanders, V. P.; Kogan, E.I.; Research of direct-flow cyclone separators of system of an ash collection of a state district power station.*Heat Power Eng*.**1968**,*8,* 45–48, (in Russian).
9. Kutateladze, S.S.; Gyroscopes, E.P.; and Terekhov, V. I; Aerodynamics and тепломассообмен in the Restricted Vortex Flows. Novosibirsk: An Academy of Sciences of the USSR;**1987**.
10. Kutateladze, S.S.; and Styrikovich, M. A; Hydraulics of Gazo-Liquid Systems. State Power Publishing House;**1958**.
11. Levich, V.G.; Physical and Chemical Hydrodynamics. Fizmatgiz;**1959**.
12. Potapov, I. P.;and Kropp, L.D.; Batarejnye Cyclone Separators. M.: Energy;**1977**.
13. Prokofichev, N.N.; Reznik, V. A; Aleksandrovich, E.I.; and Yermolaev, V.V.; To sampling of the ash collector for coppers of industrial and municipal power engineering. *Power Eng*.**1997**,*8,* 12–13, (in Russian).
14. Pirumov, A.I.; Aerodynamic Bases of the Inertia Separation. M.: A State Power Publishing House;**1961**.
15. Straus, V.; Industrial Clearing of Gases. Moscow:Chemistry; **1981**.
16. The Directory after a Heat—and to an Ash Collection. Eds. Birger, M. I.; Valdberg, A. J.; Mjagkov, B.I.; et al. Under Rusanova's, A. A.; General Edition. Atom Publishing House;**1983**.
17. Smulsky, I.I.; Aerodynamics and Processes in Whirlpool Chambers. Novosibirsk: "Science";**1992**.
18. Saburov, E.N.; Aerodynamics and Convective Heat Transfer in Cyclonic Heating Installations. Publishing House Leningrad;**1982**.
19. Saburov, E.N.; Carps, S.V.;and Ostashev, S.I.; Heat Exchange and Aerodynamics of the Twirled Stream in Cyclonic Devices. Leningrad: Publishing House;**1989**.
20. Sedov, L.I.; Similarity Methods and Dimensions of a Quantity in the Mechanic. State Power Publishing House;**1954**.

21. Shtokman, E.A.; Air Purification. M.: Publishing House ASV;**1999**.
22. Fraser, R. P.; Sixth Symposium (International) on Combustion. New York, London; **1957.**
23. Giffenen, E.;and Lamb, A. J.; The effect of air density on spray atomization.*The Motor Ind. Res. Ass.* Report no 5, **1953**.
24. Pohlhausen, K.;and Zeitschr, F.;Angew. Math, und Mech.**1921**, *1*.
25. Tate, R. W.;and Marschall, W. R.; Atomization by centrifugal pressure nozzles.*Chem. Eng. Progress.***1953**, *49(4a)*, 5.
26. Taylor, G. I.;*Proc. of the 7-th Int. Congress Appl. Mech.***1948**, *2*,Part 1.
27. Taylor,G. I.;*The Quart. J. Mech. Appl. Math.***1950**,*3*,Part 2.
28. Turner, G. M.;and Mouton, R. W.; Drop size distribution from spray nozzles.*Chem. Eng. Progress.***1953**,*49(4)*.
29. Tresch; Chemie Eng. Technik.**1954**, *26(6)*.
30. Nuкiyama, S.;and Tanasawa, J.; Experiments on the atomisation of liquids in an air stream.*Rep. 1 Trans. Trans. Soc. Mech. Eng. (Japan).***1938**, *4(14)*.
31. Nukiyama, S.;and Tanasawa, J.; Experiments on the atomisation of liquids in an air stream. *Rep. 4Trans. Trans. Soc. Mech. Eng. (Japan).***1938**,*5(18)*.
32. Radcliffe, A.;and Clare, H.; Rep. NR 144 British NGTE.**1953.**
33. Miesse C. C.; Correlation of experimental data on the disintegration of liquid jets.*Ind. Eng. Chem.***1955**, *47(9)*.
34. Probert, R. P.; The influence of spray particle size and distribution in the combustion of oil droplets.*Phil. Mag.* **1946**, *37(265)*.
35. Woltjen, A.; Ober die Feinheit der Brennstoff-Zerstaubung. Darmstadt;**1925.**

CHAPTER 7

DERIVATION OF AN EQUATION OF TRAFFIC OF DISPERSION PARTICLES AND CALCULATION OF FRACTIONAL EFFICIENCY OF CLEARING OF GAS

CONTENTS

7.1 INTRODUCTION

Theoretical and to an experimental research of multiphase turbulent flows books Zhou (1993) are devoted, to Volkova et al. (1994), Gorbis and Spo-koyny (1995), Crowe et al. (1998), Varaksina (2003) and surveys Eaton and Fessler (1994), Elghobashi (1994), McLaughlin (1994), Crowe et al. (1996), Simonin (1996), Zajchika and Pershukova (1996), Loth (2000), Sommerfeld (2000), Mashayek and Pandya (2003). In these books and surveys many questions connected with hydrodynamics and heat exchange of turbulent flows are taken up. In spite of the fact that the first work under the theory of dispersion turbulent flows has appeared rather for a long time (Barenblatt, 1953), intensive development of this area of mechanics and heat exchange was initiated only last 20 years. The basic theoretical problems originating at modelling of two-phase dispersion turbulent flows in comparison with monophase, are connected with following physical processes: interacting of corpuscles (drops, vials) with turbulent whirlwinds of a continuous phase; interacting of corpuscles with each other as a result of collisions; changes of phase, concretion or crushing; effect turbulent fluctuations for speed of changes of phase; interacting of corpuscles with a surface restricting a stream and sedimentation; return effect of corpuscles on turbulence; a dispersion, accumulation and fluctuations of concentration of corpuscles. The most exact and detailed information on formation of a turbulent multiphase stream can be gained on the basis of application of a method of direct numerical modelling *(DNS)* for the bearing continuous medium in a combination to stochastic approach Lagrange for a dispersoid. At direct numerical modelling all spectrum of turbulent whirlwinds, including fine-scale, responsible for a dissipation of energy of turbulence is presented [1–25].

However *DNS* demands the big expenses of a time even at attraction of the most rapid-transfer computers and consequently it is used, mainly, as numerical experiment for testing or gauging of more economic methods of calculation of turbulent flows [26–32]. In a method of large whirlwinds *(LES)* direct modelling only the large whirlwinds which space scale exceeds a size of a numerical grid is made, and fine-scale fashions appear out of limits of solvability and are presented by a semiempirical way. *LES* it is applicable for modelling of behaviour of corpuscles, a time of which dynamic relaxation much more than time microscale of turbulence, than in the theory of monophase turbulence (Armenio et al.,1999; Boivin et al.,2000; Yamamoto, 2001; Kuerten and Vreman, 2005; Fede and Simonin, 2006). In the capacity of a key factor of adequacy of the presented models are accepted the consent with known results from the literature of numerical modelling on the basis of *DNS* or *LES* for a continuous phase in a combination with Lagrange trajectory a method for a dispersoid. Such approach gives the powerful tool for verification of models as numerical experiments unlike the physical allow to gate out the investigated phenomenon in "a pure" aspect without effect of complicating and distorting factors (Zaichik, 2005; 2007).

7.2 STATEMENT OF PROBLEM, ASSUMPTIONS

For a statement of problem of modeling and the subsequent research of the processes proceeding in whirlwind centrifugal apparatuses, it is necessary to define association between parameters of the twisting device and a current formed by it. As numerical modeling of three-dimensional currents for today is problematic, the given problem merges with a known problem of characteristics of the twirled currents and twisting devices [33–42].

Let's observe the mechanism of clearing of gas emissions on an instance of a dynamic spray scrubber.

Dust trapping in a scrubber is based on use of a centrifugal force. The gaseous-dust stream with a great speed on a tangent arrives in a cylindrical part of the body and executes a motion on a descending spiral. Under the influence of a centrifugal force originating at a rotary motion of a stream, a dust corpuscle move to apparatus walls (Figure7.1).

At motion in a twirled curvilinear gas stream of a corpuscle of a dust are under the influence of a centrifugal force and force of resistance [43–50].

The analysis of the twirled gas-dispersed current in a scrubber we will spend at following assumptions:

1. Gas is considered ideal and incompressible liquid, hence, its motion potentially.
2. The gas stream is axisymmetric and stationary.
3. The peripheral projection of speed of gas changes in a manner
$$w_\varphi = const \cdot \sqrt{r}$$

This law observed in experiments (Barchatenko, 1974), will allow to gain the simple solution convenient for quantitative analysis of motion of corpuscles.

3. The corpuscle does not change the form and diameter in a time, does not occur neither its crushing, nor concretion. The deviation of the form of a corpuscle from sphere is considered by factor to.

4. The flow of a corpuscle a gas stream has viscous character. Turbulent pulsations of gas are not considered that will be co-ordinated with deductions of work (Lagutkin, 2004): the turbulent diffusion of corpuscles does not render appreciable effect on dust separation process.

5. Forces Zhukovsky, Archimedes, weight as the specified forces on some usages it is less in comparison with forces of an aerodynamic resistance and centrifugal (Starchenko, 1999) are not considered.

6. Concentration of a dust is small; hence, it is possible not to consider interacting of corpuscles.

7. We neglect non-uniform distribution of an axial projection of speed of gas on radius that is according to the data of work (Shilaev, 2003) according to which the axial projection of speed of corpuscles poorly changes on pipe radius.

Because of twirl of a cleared stream in a scrubber the field of the inertia forces which leads to separation of a mix of gases and corpuscles is created. Therefore for calculation of mechanical trajectories of corpuscles it is necessary to know their equations of motion and aeromechanics of a gas stream. According to the assumption of small concentration of a dust, effect of corpuscles on a gas stream it is possible to neglect. Hence, it is possible to observe motion of a separate corpuscle in the field of speeds of a gas stream.

At motion in a twirled curvilinear gas stream of a corpuscle of a dust are under the influence of the forces presented on (Figure 7.1).

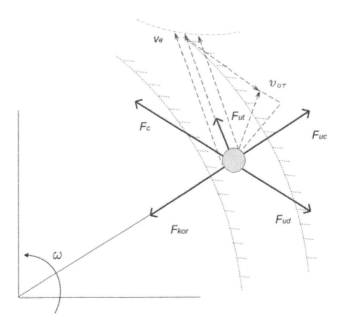

FIGURE 7.1   The forces acting on a corpuscle in a gas stream.

The problem of definition of paths of corpuscles in a scrubber decomposes on two:
- Definition of a field of speeds of a gas stream,
- Integration of the equations of motion of a corpuscle in terms of a design field of speeds of gas.

The assumption about a rotational symmetry of an observed problem (except for an inlet opening) allows using by consideration of motion of corpuscles a cylindrical coordinate system.

The greatest difficulty is represented by trapping of a finely divided dust for which force of resistance with sufficient accuracy is computed by formula Stokes, these reasons cause the third assumption. At increase in a dustiness the factor of clearing of the apparatus grows (Starchenko, 1999), therefore calculation of parameters of a scrubber at a small dustiness, (according to an assumption 6), secures its minimum efficiency.

7.3 DEVELOPMENT OF MOTION OF A CORPUSCLE

For calculation of paths of corpuscles it is necessary to know their equations of motion. Such problem for some special case dares the author (Brzik, 2000).

Let's inject co-ordinate system OXYZ. Its axis OZ we will direct along an axis of symmetry of a scrubber (Figure 7.2 see). The law of motion of a mote in motionless co-ordinate system OXYZ can be written down in a following aspect:

$$m\frac{d\overrightarrow{v}}{dt} = \overrightarrow{F_{Cm}} \tag{7.1}$$

where m—weight of a corpuscle;

dv_p—speed of a corpuscle;

F_{st}—an air force (it is computed by formula Stokes).

For calculations it is necessary to present a vector Eq. (7.1) motions of a corpuscle in the scalar form.

Corpuscle rule we will set in its cylindrical co-ordinates (r; φ; z). Speed of a corpuscle it is definable three составляющими: $U_ч$—tangential, $V_ч$—radially extending; $W_ч$—axial speeds.

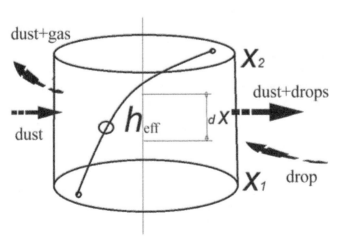

FIGURE 7.2 Vector of a traverse speed of a corpuscle.

Let us accept coordinate system $O'X'Y'Z'$ let axis $O'X'$ passes through a corpuscle, and axis $O'Z'$ lies on axis OZ. The accepted frame of reference moves is forward on axis OZ with a speed $W_ц$ and is twirled round it with angular speed

$$\omega(t) = \frac{U_p}{r_p} \qquad (7.2)$$

The equation of motion of a corpuscle in weight $m = \frac{1}{6}\pi\rho_+ d_+^3$
In coordinate system $O'X'Y'Z'$ will become:

$$m\frac{d\vec{v_ц}}{dt} = \vec{F_{st}} - m\vec{a_0} + m\left[\vec{r_ц} \cdot \vec{\omega}\right] + m\left[\vec{r_ц} \cdot \vec{\omega}\right] + m\left[\vec{\omega} \cdot \left[\vec{r_ц} \cdot \vec{\omega}\right]\right] + 2m\left[\vec{v_ц} \cdot \vec{\omega}\right]$$

$$\frac{d\vec{v_ц}}{dt} = \frac{1}{m}\vec{F_{st}} - \vec{a_0} + \left[\vec{r_ц} \cdot \vec{\omega}\right] + \left[\vec{\omega} \cdot \left[\vec{r_ц} \cdot \vec{\omega}\right]\right] + 2\left[\vec{v_ц} \cdot \vec{\omega}\right] \qquad (7.3)$$

where a_0—a vector of forward acceleration of a frame of reference;
dv_p—speed of a corpuscle;
r_p—a corpuscle radius-vector;
$\left[\vec{r_ц} \cdot \vec{\omega}\right]$—The acceleration caused by irregularity of twirl;
$\left[\vec{\omega} \cdot \left[\vec{r_ц} \cdot \vec{\omega}\right]\right]$—A centrifugal acceleration;
$2\left[\vec{v_ц} \cdot \vec{\omega}\right]$—Coriolis acceleration.
First composed the right side of an Eq. (7.3) represents the force acting with a gas stream on a corpuscle, and it is defined by formula Stokes:

$$F_{st} = 3\pi\mu_g d_p\left[\vec{v_g} - \vec{v_p}\right] \qquad (7.4)$$

where μ_g—dynamic viscosity of gas.
The another composed Eq. (7.3) is defined as

$$\frac{dW_ц}{dt}\vec{e_z} = \frac{dW_ц}{dt}\vec{e_{z'}} \qquad (7.5)$$

restructure the summand:

$$\left[\vec{r_u} \cdot \vec{\omega}\right] = \left[\vec{r_u} \cdot \frac{d\vec{\omega}}{dt}\right] = \left[\vec{r_u} \cdot \frac{d}{dt}\left(\frac{U_u}{r_u}\vec{e_z}\right)\right] = -r_u\left(\frac{1}{r_u}\frac{dU_u}{dt} - \frac{U_u}{r_u^2}V_x\right)\vec{e_z} = -\left(\frac{dU_u}{dt} + \frac{U_uV_u}{r_u}\right)\vec{e_y}$$

$$\left[\vec{\omega} \cdot \left[\vec{r_u^{-1}} \cdot \vec{\omega}\right]\right] = \frac{U_u^2}{r_x}\left[\vec{e_z} \cdot \left[\vec{e_x} \cdot \vec{e_z}\right]\right] = -\frac{U_u^2}{r_u}\left[\vec{e_z} \cdot \vec{e_v}\right] = \frac{U_u^2}{r_u}\vec{e_x} \tag{7.6}$$

$$2\left[\vec{v_u} \cdot \vec{\omega}\right] = 2v_{x'}\left[\vec{e_{x'}} \cdot \vec{\omega}\right] = 2v_{x'}\frac{U_x}{r_u}\left[\vec{e_{x'}} \cdot \vec{e_{z'}}\right] = \left(-2\frac{U_uV_u}{r_u}\right)\vec{e_y}$$

where— $\vec{e_u}, \vec{e_v}, \vec{e_z}$ Crosscuts of a system of reference also it is considered that

$$\vec{r_u} \quad \vec{e_x} \quad r_u \quad v_x \quad V_u$$

Substituting these expressions in the equation of motion Eq. (7.3) it is gained

$$m\frac{d\vec{v_u}}{dt} = \vec{F_{Cm}} - m\vec{a_0} + m\left[\vec{r_u} \cdot \vec{\omega}\right] + m\left[\vec{r_u} \cdot \vec{\omega}\right] + m\left[\vec{\omega} \cdot \left[\vec{r_u} \cdot \vec{\omega}\right]\right] + 2m\left[\vec{v_u} \cdot \vec{\omega}\right] \tag{7.7}$$

$$\frac{d\vec{v_u}}{dt} = \frac{1}{m}\vec{F_{Cm}} - \vec{a_0} + \left[\vec{r_u} \cdot \vec{\omega}\right] + \left[\vec{\omega} \cdot \left[\vec{r_u} \cdot \vec{\omega}\right]\right] + 2\left[\vec{v_u} \cdot \vec{\omega}\right]$$

The equation in projections to axes of co-ordinate system O'X

$$\begin{cases} \dfrac{dU_x}{dt} = \dfrac{3}{4} \cdot \dfrac{\rho_g}{\rho_p} \cdot \dfrac{\xi_p}{d_p} \cdot U_{tm} \cdot (V_\partial - U_\partial) + g; U_x = \dfrac{dx}{dt} \\[3mm] \dfrac{dU_\phi}{dt} = \dfrac{3}{4} \cdot \dfrac{\rho_g}{\rho_p} \cdot \dfrac{\xi_p}{d_p} \cdot U_{tm} \cdot (V_\phi - U_\phi) + \dfrac{\omega \cdot r}{t}; U_\phi = \omega \cdot r = r\dfrac{d\phi}{dt} \\[3mm] \dfrac{dU_r}{dt} = \dfrac{3}{4} \cdot \dfrac{\rho_g}{\rho_p} \cdot \dfrac{\xi_p}{d_p} \cdot U_{tm} \cdot (V_r - U_r) - \dfrac{\omega \cdot r}{t} + \omega^2 r; U_r = \dfrac{dr}{dt} \end{cases} \tag{7.8}$$

We have gained the equation of motion of a corpuscle in a twirled gas stream in projections to axes of a cylindrical coordinate system.

Analogously the system of the equations of motion of drops of a liquid registers:

$$
\begin{cases}
\dfrac{d\,W_x}{dt} = \dfrac{3}{4}\cdot\dfrac{\rho_g}{\rho_l}\cdot\dfrac{\xi_\kappa}{d_\kappa}\cdot W_{otn}\cdot(V_x - W_x) + g; W_x = \dfrac{dx}{dt} \\[2mm]
\dfrac{d\,W_\varphi}{dt} = \dfrac{3}{4}\cdot\dfrac{\rho_g}{\rho_l}\cdot\dfrac{\xi_\kappa}{d_\kappa}\cdot W_{otn}\cdot(V_\varphi - W_\varphi) + \dfrac{\omega\cdot r}{t}; W_\varphi = \omega\cdot r = r\dfrac{d\varphi}{dt} \\[2mm]
\dfrac{d\,W_r}{dt} = \dfrac{3}{4}\cdot\dfrac{\rho_g}{\rho_l}\cdot\dfrac{\xi_\kappa}{d_\kappa}\cdot W_{otn}\cdot(V_r - W_r) - \dfrac{\omega\cdot r}{t} + \omega^2 r; W_r = \dfrac{dr}{dt}
\end{cases}
\tag{7.9}
$$

From system of the equations of motion of a corpuscle follows that its path depends on following factors: d_p—diameter of corpuscles (drops d_d); ρ_p—corpuscle density (liquids ρ_l); μ_g—dynamic viscosity of gas (a liquid μ_l); r—radius (a characteristic size) apparatus; U_{go}—initial tangential speed of gas, and also from air swirler geometry α, Ď, ž.

The approach which is based on the theory of geometrical similitude of dedusters is offered, and allowing to consider a compulsory twisting of a stream and air swirler critical bucklings.

In the capacity of a similarity parameter following dimensionless parameters are offered: relative diameter of corpuscles d_p/D_0; a relative density of corpuscles ρ_p/ρ_g; Reynolds number $Re = v_p d_p/v_g$; Strouhal number $Sh = \omega\,D_0/v_g$, considering angular speed of a rotor, a characteristic size of air swirler R_z. In number of diagnostic variables are not incorporated, as not having usually essential value, acceleration of free falling and the parameters characterizing compressibility of gas and thermal processes in it. The gas dustiness in the assumption of independent motion of corpuscles of a dust also is not included into number of diagnostic variables.

If to inject factor R_z (a characteristic size of an air swirler) the system of the differential Eqs. (7.8–7.9) can be transcribed so:

$$
\begin{cases}
\dfrac{dU_x}{dt} = \dfrac{18\mu_g}{\rho_p d_p^2}\left(\dfrac{R_0}{r_p}U_{g0} - U_p\right) - \dfrac{U_x V_x}{r_q} + g \\[2mm]
\dfrac{dU_\varphi}{dt} = \dfrac{18\mu_g}{\rho_p d_p^2}\left(\dfrac{Q_0}{R_0^2}\cdot V'\left(\dfrac{\alpha}{R_z};\dfrac{z}{R_z};\dfrac{r}{R_z};\dfrac{D}{R_z};...\right) - V_\varphi\right) + \dfrac{U_\varphi^2}{r_q} + \dfrac{\omega\cdot r}{t} \\[2mm]
\dfrac{dU_r}{dt} = \dfrac{18\mu_g}{\rho_p d_p^2}\left(\dfrac{Q_0}{R_0^2}\cdot W'_g\left(\dfrac{\alpha}{R_z};\dfrac{z}{R_z};\dfrac{r}{R_z};\dfrac{D}{R_z};...\right) - V_r\right) - \dfrac{\omega\cdot r}{t} + \omega^2 r
\end{cases}
\tag{7.10}
$$

Let's inject dimensionless cylindrical co-ordinates:

$$r' = \frac{r}{R_0}, \; z' = \frac{z}{R_0} \; \varphi' = \varphi$$

Real speeds it is renewable the dimensionless:

$$U_x' = \frac{R_z^2}{Q_0} \cdot U_x, \; U_\varphi' = \frac{R_z^2}{Q_0} \cdot U_\varphi, \; U_r' = \frac{R_z^2}{Q_0} \cdot U_r.$$

Dimensionless time: $t' = \frac{Q_0}{R_z^3} \cdot t.$

For non-dimensional quantities definitions of speed remain:

$$U'_x = \frac{dr'}{dt'}, \; U'_\varphi = \frac{dz'}{dt'}, \; U'_r = \frac{d\varphi'}{dt'} \cdot r'.$$

Let's denote:

$$C_p = \frac{\mu_\varepsilon R_z^3}{\rho_u Q_0 d_u^2}, \; C_z = \frac{y \sin \beta}{R_2}.$$

Having made matching substitutions in system Eq. (7.10), after transformation we will gain system of the differential equations for dimensionless quantities:

$$\frac{dU'_x}{dt'} \quad 18C_p \quad \frac{C_r}{r'_u} \; U'_u \quad \frac{U'_u V'_u}{r'_u}$$

$$\frac{dU'}{dt'} \quad 18C_p \; U'_z \quad ; z' ; r' ; D'; \dots \quad V'_u \quad \frac{U'^2_u}{r'_u} \qquad (7.11)$$

$$\frac{dU'_r}{dt'} \quad 18C_p \; U'_z \quad ; z' ; r' ; D'; \dots \quad W'_u$$

At starting conditions

$$r'_u \big|_{t=0} = r'_0, \; z'_u \big|_{t=0} = z'_0, \; \varphi'_u \big|_{t=0} = 0 \text{—for cylindrical coordinates,}$$

$$U'_x \big|_{t=0} U'_{z0}, \; U'_\varphi \big|_{t=0} = U'_{z0}, \; U'_r \big|_{t=0} = U'_{z0}. \text{—for speed of a corpuscle}$$

And two dimensionless parametersC_p and $C_r$solution of this system defines a corpuscle path in the cyclone separator.

The system Eq. (7.11) is system of the ordinary differential equations

The another order. Knowing radially extending $\acute{U}_r$ and axial speeds $\acute{U}_x$ a gas stream, it is possible to integrate and gain a corpuscle path.

The solution of system Eq. (7.11) at an equal aspect of functions

$$U'_\phi\left(\alpha; z'; r'; D';...\right); U'_r\left(\alpha; z'; r'; D';...\right)$$

I.e. at the equal form of dedusters, it is defined by dimensionless parameters of $C_p$and $C_r$which depend from regime and critical bucklings. At equality of these parameters the corpuscle path will be equal. Thus, $C_p$and C_r are a similarity parameter for motion of corpuscles of a dust in a scrubber. Identity of functions is necessary for full similitude of motion of corpuscles also

$$U'_\phi\left(\alpha; z'; r'; D';...\right); U'_r\left(\alpha; z'; r'; D';...\right)$$ (Geometrical similarity).

It is visible that the criterion $C_p$is defined by physical parametres of a dust-laden gas mix and "specific productivity of a scrubber. The criterion$C_g$is defined by sizes of a conic air swirler and a slope of its guide vanes on which the relationship between tangential and radial speeds, and consequently, a relationship of an inertial force and force of resistance depends.

7.4 SAMPLING OF A METHOD OF THE SOLUTION OF THE EQUATIONS OF MOTION OF A CORPUSCLE

The motion equations, also as well as the equation of the Laplace, dare numerically.

Let's note that the system of the equations of motion of a corpuscle is rigid as contains sharply excellent values of derivatives (Elghobashi, 1983). The matrix of system of the equations of the first oncoming for the equations of motion looks like:

$$\begin{pmatrix} -18K_t - \dfrac{V}{r} & -\dfrac{U}{r} & 0 & -18\dfrac{K_t K_v}{r^2} - \dfrac{UV}{r^2} & 0 \\[2mm] \dfrac{2U}{r} & -18K_t & 0 & 18K_t\dfrac{dV_g}{dr} - \dfrac{U^2}{r^2} & 18K_t\dfrac{dV_g}{dz} \\[2mm] 0 & 0 & -18K_t & 18K_t\dfrac{dW_g}{dr} & 18K_t\dfrac{dW_g}{dz} \\[2mm] 0 & 1 & 0 & 0 & 0 \\[1mm] 0 & 0 & 1 & 0 & 0 \end{pmatrix}. \quad (7.12)$$

Its number of conditionality nearby 10^5 that speaks about rigidity of system of the equations of motion.

Use of obvious methods of the solution of system Eq. (7.12) imposes strong restrictions on magnitude of an integration step because of possibly unstable stability of computing process. To solve this problem it is possible by application of implicit methods (Crowe, 1996). However in this case on each iteration on a time it is necessary to decide the nonlinear equations that considerably complicates calculation. Therefore it has been decided to use a semi-implicit method of the solution of system of the equations of motion. In an applied method each design difference equation is implicit on that variable on which differentiation is made, and on other variables of the equation are obvious. Now at integration of rigid systems of the equations the Gear method (Fede, 2006) is widely used.

For integration of the equations of motion of a corpuscle Eq. (7.11) have been tested a Gear method of the first order (which coincides with Euler's implicit method) and a Gear method of the other order. If to mark out value of a variable on n–м an integration step as x^n, and on $(n+1)$- th as x^{n+1} the difference analogue Eq. (7.11) at integration by Euler's semi-implicit method will look like (in terms of Eq. (7.12)):

$$
\left\{
\begin{array}{l}
\dfrac{U^{n+1}-U^n}{\Delta t}=18K_t\left(\dfrac{K_v}{r^n}-U^{n+1}\right)-\dfrac{U^{n+1}V^n}{r^n} \\[3ex]
\dfrac{V^{n+1}-V^n}{\Delta t}=18K_t\left(V_g\left(r^n,z^n\right)-V^{n+1}\right)+\dfrac{\left(U^n\right)^2}{r^n} \\[3ex]
\dfrac{W^{n+1}-W^n}{\Delta t}=18K_t\left(W_g\left(r^n,z^n\right)-W^{n+1}\right) \\[3ex]
\dfrac{r^{n+1}-r^n}{\Delta t}=V^{n+1} \\[3ex]
\dfrac{z^{n+1}-z^n}{\Delta t}=W^{n+1} \\[3ex]
\dfrac{\phi^{n+1}-\phi^n}{\Delta t}=\dfrac{U^{n+1}}{r^n}
\end{array}
\right.
$$

$$(7.13)$$

(Subscripts and primes are hauled down).

It is necessary to note that application of a semi-implicit method, at conservation of simplicity of scalings, secures with much wider limits of stability of computing process, rather than obvious methods. So the overhead basil of a resistant to integration step on a time increases by two order in case of application of a method of Euler.

Euler's semi-implicit method secures with sufficient stability, accuracy and speed of scalings at simple program of implementation and consequently it is applied to integration of the equations of motion. The Gear method of the other order has slightly best speed of scalings (at given accuracy), but it is more difficult at program implementation. Conditional stability of a semi-implicit method of integration imposes restrictions on magnitude of an integration step, thus the overhead basil of a step of resistant to integration of a method of Euler secures with necessary accuracy. Application of a semi-implicit Gear method of the another order does not allow to improve essentially stability of computing process, therefore its application for increase in accuracy of calculations is inexpedient.

At calculation of a path of corpuscles their speed is very non-uniform, therefore theintegration step is chosen automatically depending on speed of a corpuscle. According to Eq. (7.13) it is possible to write down the following iterative formula for the numerical solution of the equations of motion and construction of mechanical trajectories of corpuscles of a dust in the separator:

$$\begin{cases} U^{n+1} = \left(U^n + 18K_t \Delta t^n / r^n\right) \big/ \left(1 + \left(18K_t + V^n / r^n\right)\Delta t^n\right) \\ V^n = \left(V^n + \left(18K_t V_g \left(r^n, z^n\right) + \left(U^n\right)^2 / r^n\right)\Delta t^n\right) \big/ \left(1 + 18K_t \Delta t^n\right) \\ W^n = \left(W^n + 18K_t W_g \left(r^n, z^n\right)\Delta t^n\right) \big/ \left(1 + 18K_t \Delta t^n\right) \\ r^{n+1} = r^n + V^{n+1}\Delta t^n \\ z^{n+1} = z^n + W^{n+1}\Delta t^n \\ \phi^{n+1} = \phi^n + U^{n+1}\Delta t^n / r^{n+1} \\ \Delta t^{n+1} = C \big/ \sqrt{\left(V^n\right)^2 + \left(W^n\right)^2} \end{cases}$$

$$(7.14)$$

This iterative formula is expression of a semiimplicit method of Euler

7.5   ESTIMATION OF ACCURACY OF SCALINGS

All scalings should be made with a split-hair accuracy, therefore a sub-interval and integrations is chosen small, and quantity of iterations—big.

The lapse is admitted at several stages of calculation:
- At replacement of the equation of the Laplace with finite-difference equations;
- At the solution of system of finite-difference equations;
- At integration of the equations of motion;
And on some other (they are less essential).

The lapse of the solution of the equation of the Laplace collects at integration of the equations of motion. To consider lapses and their interactions,it is separately inconvenient, therefore the solution lapse is designed not at the intermediate stages, and by definitive result.

The design program builds paths of corpuscles which depend not only on the set parameters of process of a dust separation, but also from calculation parameters:

- h—a subinterval of area S in MCR;
- N—numbers of iterations at the solution of system of finite-difference equations;
- Δt—an integration step of the equations of motion.

Thus, the corpuscle radius-vector depends not only on parameters of the cyclone separator, a dust and gas, but also from calculation parameters:

$$\vec{r} = \vec{r}\left(t, d_{_q}, \rho_{_q}, Q_0, ..., h, N, \Delta t, ...\right).$$

At toughening of parameters of calculation $h{\to}0$, $N{\to}\infty$, $\Delta t {\to}0$

The finishing point of a path of a corpuscle is aimed to the limiting value, therefore for a lapse measure it is possible to accept $\Delta \dot{r}$—distance between finishing points of the limiting and approached paths.

To size up a lapse of the numerical solution it is possible on magnitude of differential of a radius-vector of a finishing point of a path of a corpuscle, considering as variables design variables. Other arguments of function $\dot{r}$ are fasted for the most adverse for case calculation.

The final decision lapse was sized up by formula:

$$\varepsilon_r = \left|\frac{\partial r}{\partial\left(1/N\right)}\right|\frac{1}{Nr} + \left|\frac{\partial r}{\partial h}\right|\frac{h}{r} + \left|\frac{\partial r}{\partial \Delta t}\right|\frac{\Delta t}{r}. \tag{7.15}$$

7.6 MATHEMATICAL MODEL OF PROCESS OF A DUST SEPARATION IN THE APPARATUS

In the conditions of cross motion and interacting of a dusty gas stream and drops of an irrigating liquid sedimentation of firm and liquid corpuscles on drops of the sprayed liquid occurs by means of three mechanisms: the inertia sedimentation, interception of corpuscles and diffused sedimentation (Crowe, 1995). At clearing of gases of finely divided corpuscles when the characteristic size of corpuscles of a dispersoid changes within 5–150 µ, and diameter of drops of an irrigating liquid makes about 300 µ(Sommerfeld, 2003), the contribution of mechanisms of interception of corpuscles and diffused sedimentation appears insignificant and efficiency of process clearings of gas emissions is almost completely defined by the mechanism of the inertia collision of corpuscles with drops of an irrigating liquid.

On the given mechanism the corpuscle, thanks to the weight, possesses sufficient inertia to move rectilinearly on a heading to a trapping body-target, re-cutting stream-lines of a current of a bearing phase. Efficiency of capture at the inertia sedimentation is defined by share of the corpuscles in regular intervals distributed in a gas stream which can be trapped the sphere which cross-sectional area is equal to a body-target frontal area. Calculation of efficiency of sedimentation of corpuscles at the inertia collision with sphere is usually carried out on empirical dependences.

For an estimation of efficiency of the inertia separation of corpuscles of a dispersoid on drops of liquid in a working zone of a scrubber the transport process of weight of a dispersoid from a gas stream in a liquid in a zone of a flat radial torch has been observed. The analysis was spent in the cylindrical co-ordinate system rigidly connected with an axis of a scrubber. In spite of the fact that the density of drops in a liquid torch is high enough, changes of speed of a gas stream in a torch was neglected. Was considered that displacement of drops in a vertical heading under the influence of a gas stream and a gravity is insignificant. It was assumed also that corpuscles of a dispersoid a move without slippage concerning a gas stream.

In a zone of cross interacting of phases (Figure 7.3) has been gated out a volume element (torus) with the cross-section parties dz and dr and change of quantity of a dispersoid in a gas stream is defined at passage of the given volume by it

$$dM = -2\pi W_z \frac{dc}{c_z} r dr dz \qquad (7.16)$$

where W_z—speed of a gas stream; c—concentration of a dispersoid in gas.

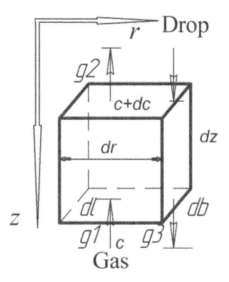

FIGURE 7.3   The circuit design of interacting of streams.

Change of quantity of a dispersoid in a gas stream at volume element passage is caused by capture of corpuscles by drops of liquid and defined by the total square of cross-section of drops dF_k, efficiency of capture of corpuscles a single drop η_k and speed of relative motion of corpuscles and drops

$$dM = \eta_k V_{otn} c dF_k \qquad (7.17)$$

At calculation of magnitude of relative speed of a corpuscle it was considered that in the conditions of a cross current the radial velocity of drops V_r, considerably surpasses in magnitude speed of a gas stream (Sommerfeld, 2000).

$$V_{otn} = \sqrt{W_z^2 + V_r^2} \approx V_r \qquad (7.18)$$

The total square of cross-section of drops is connected with weight of drops in the gated out elementary volume

$$dF_k = \frac{3}{2\rho_k a} dM \tag{7.19}$$

The weight of drops in an elementary volume was defined through magnitude of the resulted speed of a liquid phase in torch U and a time of motion of drops in an elementary volume $d\tau$

$$dM = 2\pi\rho_k Urdzdr \tag{7.20}$$

where $U = \dfrac{L}{2}\rho_k \pi r b, \ d\tau = \dfrac{dr}{V_r}$

L—A liquid mass flow in a spray; a b-thickness of a drop layer.

By comparison of expressions (7.17) and (7.20) the equation characterizing change of concentration of a dispersoid on a thickness of a spray of a liquid has been gained

$$\frac{1 \cdot dc}{c \cdot dz} = -B \tag{7.21}$$

where

$$B = \frac{3}{4} \cdot \frac{\rho_g}{\rho_l} \cdot \frac{L}{G} \left(\frac{(1-\chi) \cdot \varepsilon}{\chi_k} \right) \cdot \eta_K \cdot \frac{1-m^2}{b \cdot d_K} \cdot \frac{t_1}{t_1^2 - m^2} \cdot \frac{\sin\beta}{R_2};$$

where

$$m = \frac{R_1}{R_2} \quad t_1 = \frac{y \cdot \sin\beta}{R_2}$$

Where G—a gas rate, kg/m³; L—the liquid charge, kg/m³; b—a thickness of a drop layer, m; β—an angle of a cone of an air swirler; R_1, R_2—radiuses of the basis of a conic air swirler; ρ_g, ρ_l—gas and liquid density, kg/m³; ε—factor of loss of blade twist; χ—parametre of a drop layer; relaxation time of corpuscles of a dust from a relationship:

$$\frac{1}{A} = \frac{\delta_p^2 \cdot \rho_p}{18 \cdot \rho_g \cdot V_g}$$

Integration of the differential Eq. (7.21) at starting conditions

$$z=z_n; \; c = c_{_H}(7.22)$$

The following dependence for calculation of concentration of corpuscles of a dispersoid in a gas stream has been gained

$$C_{XK} = C_{XH} \cdot \exp\left(-B \cdot (z - z_n)\right)$$

Also concentration of corpuscles in gas on an exit from a drop layer is defined

$$C_{XK} = C_{XH} \cdot \exp\left(-B \cdot b\right) \tag{7.23}$$

From the formula (7.23) follows, what even in case of a dispersoid uniform distribution in a gas stream on an entry in a spray of an irrigating liquid $(c_{_H} = const)$, concentration of an impurity in gas on an exit from a drop layer will be a variable on scrubber radius. This fact is caused, on the one hand, by expansion of a flat spray in process of removal from a source and, accordingly, water concentration decrease, and with another—falling of a radial velocity of drops of a liquid at the expense of force of an aerodynamic resistance (Mashayek, 1997; 2003; and 2002).

It has been assumed that in space between layers there is a mixing of a gas stream that leads to leveling of concentration on apparatus cross-section. Value of average concentration has been defined from the material balance equation

$$C_y = C_{yH} + 2\frac{G}{L} \cdot \frac{\rho_l}{\rho_{lg}} \cdot \frac{C_{XH}}{R_2^2 - R_1^2} \int\limits_{R_1/\sin\beta}^{R_2/\sin\beta} y\left(1 - e^{-B \cdot b}\right) dy \tag{7.24}$$

where $\eta_K = (C_H - C_k)/C_H$ —efficiency of a dust separation;

R_1—Radius of the big cone basis of an air swirler; R_2—radius of a smaller cone basis of an air swirler.

For calculation of fractional efficiency of a dust separation in a scrubber the mathematical model of a mass transport firm of gas in liquid in a zone of motion of a twirled layer of drops is made. At model formulation it was considered that the dust separation in the apparatus occurs as at the expense of centrifugal separation in the twirled stream, and by their inertia sedimentation on drops. And for small corpuscles ($\delta \leq 5$ μ) it is the basic mechanism of sedimentation.

In the assumption that the gas stream moves through a drop layer in a regime of ideal replacement, and change of quantity of a solid phase in a gas stream is caused only by the inertia capture of corpuscles of a dust by drops of liquid, dependence for calculation of concentration of corpuscles in a gas stream on an exit from a layer is gained:

$$C_{XK} = C_{XH} \cdot \exp(-B \cdot b)$$

Definition of efficiency of a deduster as the share of corpuscles which are carried away by a liquid from total of a dust, arriving with gas:

$$h_u = \frac{L \times r_g \left(C_{yk} - C_{arr}\right)}{G \times r_l \times C_{XH}} \qquad (7.25)$$

has allowed to gain the equation for calculation of efficiency of the inertia sedimentation

$$\eta_u = \frac{2}{1-m^2} \int_m^1 t_1 \left(1 - e^{-T}\right) dt \qquad (7.26)$$

where

$$T = \frac{3}{4} \frac{\rho_g}{\rho_l} \frac{L}{G} \frac{(1-\chi)\varepsilon}{\chi_k} \eta_K \frac{1-m^2}{d_k} \frac{t_1}{t_1^2 - m^2} \frac{\sin \beta}{R_2}$$

where u_{BX}—speed of gas on an entry in an air swirler, mps; u_{j0}, u_{r0}, w_j w_i, $u_{jw}u_x$—tangential and radial making speeds of gas, tangential speed of drops both tangential and radial speeds of corpuscles in a drop layer, mps; a v-thickness of a drop layer, m; d_x-average diameter of drops, m; ad_u of corpuscles, m; a =2.8 (constant)—skilled values; h=0.5 (constant) skilled values; h_x—efficiency of capture of corpuscles a single drop.

The case of a joint action of the inertia capture of corpuscles and their centrifugal separation was modeled by drops through fictitious increase in a path of motion of a corpuscle in a twirled drop layer which was sized up by parameter

$$\lambda = \sqrt{1 + \frac{U_\varphi^2}{U_x^2}} \qquad (7.27)$$

$$\eta_y = \frac{2}{1 - m^2} \int_m^1 t_1 \left(1 - e^{-\lambda \cdot T}\right) dt \qquad (7.28)$$

As one would expect, efficiency of process of the inertia separation does not depend on concentration of corpuscles in a gas phase on an entry in the apparatus. Independence of separation efficiency of gas of a thickness of a drop layer is less obvious. The given fact speaks that the quantity of a dispersoid trapped by a liquid, depends on number of drops in a layer and their surface, hence, is defined by the liquid charge, a size of drops and speed of their motion.

In actual practice a unequigranular drop layer and unequigranular composition of corpuscles in a gas stream fractional and full of efficiency of a dust separation can be designed on a principle of addition of all factors influencing separation.

7.7   CONCLUSIONS

1. The mathematical model of motion of a dispersion stream in a scrubber Is devised. At the heart of model studying of a mechanical trajectory of separate corpuscles and drops, in terms of effects regime and the design data of a scrubber which are sized up by a similarity parameter of apparatuses on the basis of similarity theory of physical processes is necessary.

2. The mathematical description of a fluid kinetics of the scrubber, allowing defining profiles an axial and peripheral component of speed of a gas stream Is offered. The model reflects essential features of the twirled motion of phases in characteristic hydrodynamic zones of the apparatus.

3. The regularity reflecting effect geometrical and operating conditions of a scrubber on magnitude of fractional efficiency on which basis the mathematical model of process of clearing of gas emissions is devised at a joint action of mechanisms of the inertia and centrifugal separation Are installed.

4. The devised model can be used at calculation and designing of apparatuses of clearing of gas emissions as relationships making it define association between technical characteristics on dedusters both their geometrical and operating conditions.

KEYWORDS

- **A velocity profile**
- **Capture of particles**
- **Fractional efficiency**
- **Mathematical model**
- **Simulation criteria**
- **The drop layer**
- **The path of corpuscles**

REFERENCES

1. The Patent 2339435 Russian Federations, Dynamic spray scrubber. Usmanova, R. R.;November 27,**2008.**
2. Barchatenko, G. M.;and Idelchik, I.E.; Effect of the form of the twisting device on a water resistance of the direct-flow cyclone separator. In: Industrial and Sanitary Clearing of Gases.**1974,** 4–7, *(in Russian).*
3. Brzik, D. A.; Automation of Calculation of Parametres of the Cyclone Separator on the Basis of Mathematical Modelling of Process of a Dust Separation. Bryansk;**2000.**
4. Varkasin, A. J.; Turbulent Flows of Gas with Firm Corpuscles. M.: Physical and Mathematical Publishing House; **2003.**
5. Wolkov, E. P.; Zajchik, L. I.;and Pershukov, V. A.; Modelling of Burning of Firm Fuel. M.: Science;**1994.**
6. Deitch, M. E.;and Fillipov, G. A.; Gas Kinetics of Two-Phase Medium. M.: Energy;**1968.**
7. Lagutkin, M. G.;and Baranov, D. A.; Estimation of act of coriolis force in apparatuses with the twirled stream. *Theoret.Bases Chem.Eng.***2004,***1,* 9–13,*(in Russian).*
8. Starchenko, A.V.; Bubenchikov, A.M.;and Burlutsky, E.S.; Mathematical model of not isothermal turbulent flow suspension in a pipe. *Thermal Phys.Aeromech.***1999,***1(6),* 59–70, (in Russian).
9. Straus, V.; Industrial Clearing of Gases. M.:Chemistry;**1981.**
10. Shilaev, M. A.; Aerodynamics and mass transfer gas-dispersed streams of studies. In: The Grant. Tomsk: Publishing House; **2003.**

11. Aref, H.;The numerical experiment in fluid mechanics II.*J. Fluids. Mech.* **1986**,*173*, 15–41.
12. Aref, H.;and Pomphrey, N.;Integrable and chaotic motions of four vortices. I. The case of identical vortices II.*Proc. R. Soc. London.* **1982**,*380A*, 359–387.
13. Bagrets, A. A.;and Bagrets, D. A.;Nonintegrability of Two Problems in Vortex Dynamics I.Chaos;**1997**,*7*, 368–375.
14. Bertozzi, A. L.;Heteroclinic orbits and chaotic dynamics in planar fluid flows II SIAM.*J. Math. Anal.***1988**,*19(6),* 1271–1294.
15. Boivin, M.; Simonin,O.;and Squires, K.D.; Direct numerical simulation of turbulence modulation by particles in isotropic turbulence. I. *Fluid Mech. 375.* **1998**,235–263.
16. Boivin, M.;and Simonin, O.D.; On the prediction of gas-solid flows with two-way coupling using large eddy simulation. *Phys. Fluids.***2000**,*12(8)*,2080–2090.
17. Campbell, L.;and Ziff, R.;A Catalog of Two-Dimensional Vortex Patterns. Los Alamos Scientific Laboratory Report No La-7384-MS. **1978.**
18. Campbell, L.J.;and Ziff, R. M.;Vortex patterns and energies in a rotation superfluidll. *Phys. Rev. B.* **1979**,*20(5),* 1886–1901.
19. Gorbis, Z.R.;and Spokoyny E. E.;Momentum and Heat Transfer in Turbulent Gas-Solid Flows. New York: Begell House Inc.; **1995.**
20. Crowe, C. T.; Sommerfeld,M.;and Tsuji, Y.;Multiphase Flows with Droplets and Particles. Florida, USA: CRC Press; **1995.**
21. Crowe, C. T.; Troutt, T. R.;and Chung, J. N.; Numerical models for two-phase turbulent flows. *Ann. Rev. Fluid Mech.* **1996**, *28*, 11–43.
22. Dhanak, M. R.;and Marshall M. P.;Motion of an elliptical vortex under applied periodic strain II.*Phys. Fluids, A.* **1993**,*5(5),* 1224–1230.
23. Elghobashi, S. E.; On predicting particle-laden turbulent flows. *Appl. Scl.Res.***1994**, 52, 309–329.
24. Elghobashi, S. E, and Abou-Arab, T. W.; A two-equation turbulence model for two-phase flows. *Phys. Fluids.***1983**, *26(4),* 931–938.
25. Elghobashi,S.;and Truesdell, G. C.; Direct simulation of particle dispersion in a decaying isotropic turbulence. *Fluid Mech.* **1992,** 242,655–700.
26. Fede, P.; Fevrier, P.;and Simonin, O.; Numerical Study of the Effect of the Fluid Turbulence Microscales on Particle Segregation and Collisions in Gas-Solid Turbulent Flows. In Proc. 5th Int. Conf. on Multiphase Flow. Paper No 343. Yokohama, Japan;**2004.**
27. Fede, P.;and Simonin, O.; Modelling of kinetic energy transfer by collision of a non-settling binary mixture of particles suspended in a turbulent homogeneous isotropic flow. In:Proc. 4th ASME-JSME Joint Fluids Eng. Conf. FEDSM2003-45735,**2003.**
28. Fede, P.;and Simonin, O.; Application of a perturbated two-Maxwellian approach for the modelling of kinetic stress transfer by collision in non-equilibrium binary mixture of inelastic particles. In:Proc. ASME Fluids Eng. Division Summer Meeting and Exhibition. Houston, USA;**2005.**
29. Fede, P.; Simonin, O.; Numerical study of the subgrid fluid turbulence effects on the statistics of heavy colliding particles. *Phys. Fluids.* **2006**, 18, 1–17.
30. Fede, P.; Simonin, O.; Villedieu Ph.; Monte Carlo simulation of colliding particles in gas-solid turbulent flows from a joint fluid-particle PDF equation. In:Proc. ASME

Joint US-European Fluids Eng. Conference. FEDSM2002-31226. Montreal, Canada;**2002.**

31. Kuerten, J.G.M.; Vreman, A. W.; Can turbophoresis be predicted by large-eddy simulation. *Phys. Fluids.* **2005,** *17* Ml, 1–4.

32. Karman von Th.;Uber den Mechanismus des Widerstands, den ein bewegter Korper in einer Fliissigkeit erfahrt II.*Gottingen Nach. Math. Phys. Kl.***1911,** 509–519.

33. Laurent-Polz, F.;Point vortices on the sphere: a case with opposite vortices II.*Nonlinearity.***2002.** *15(1),* 143–172.

34. Mashayek, F.; Jaberi, F.A.; Miller, R.S.; Dispersion and polydispersity of droplets in stationary isotropic turbulence. *Int. J. Multiphase Flow.* **1997,** *23(2),*337–355.

35. Mashayek, F.; Pandya, R.V.R.; Analytical description of particle/droplet-laden turbulent flows. *Progress Energy Combust Sci.* **2003,** *29,*329–378.

36. Mashayek F.; Taulbee, D.B. Turbulent gas-solid flows. Part II: Algebraic models. *Numer. Heat Transfer. Part B.* **2002,** *41,*31–52.

37. McLaughlin, J. B.; Aerosol particle deposition in numerically simulated channel flow. *Phys. Fluids A.***1989,***1(7),* 1211–1224.

38. McLaughlin, J.B.; Inertial migration of a small sphere in linear shear flows. *Fluid Mech.* **1991,***224,* 261–274.

39. McLaughlin, J.B.; The lift on a small sphere in wall-bounded linear shear flows. *Fluid Mech.* **1993,** *246,*249–265.

40. McLaughlin, J.B.; Numerical computation of particles-turbulence interaction. *Int. J. Multiphase Flow.* **1994,** *20(Suppl),* 211–232.

41. Simonin, O.; Deutsch, E.; Minier, J.P.; Eulerian prediction of the fluid/particle correlated motion in turbulent two-phase flows. *Appl. Sci. Res.* **1993,** *51,* 275–283.

42. Simonin, O.; Fevrier, P.; Lavieville, J.; On the spatial distribution of heavy particle velocities in turbulent flow: Turbulence. **2002,** *3(040).*

43. Sommerfeld, M.; Modelling of particle-wall collisions in confined gas-particle flows. *Int. J. Multiphase Flow.***1992,** *18(6),* 905–926.

44. Sommerfeld, M.; Inter-particle collisions in turbulent flows: a stochastic Lagrangian model. In:Turbulence and Shear Flow Phenomena — I. New York: Begell House Inc;**1999,** 265–270.

45. Sommerfeld, M.; Theoretical and Experimental Modeling of Particulate Flow: Overview and Fundamentals. Lecture Series 2000–2006. Belgium: Von Karman Institute for Fluid Dynamics;**2000.**

46. Sommerfeld, M.; Analysis of collision effects for turbulent gas-particle flow in a horizontal channel. Particle transport. *Int. J. Multiphase Flow.* **2003,** *29,* 675–699.

47. Zaichik, L.I.;and Alipchenkov, V.M.; Statistical models for predicting particle dispersion and preferential concentration in turbulent flows. *Int. J. Heat and Fluid Flow.* **2005,** *26(3),*416–430.

48. Zaichik, L.I.;and Alipchenkov, V.M.; Statistical Models of Motion of Corpuscles in a Turbulent Liquid.M.: The Physical and Mathematical Publishing House;**2007.**

49. Zhou, Y.; Wexler, A.S.;and Wang, L.P.; Modelling turbulent collision of bidisperse inertial particles. *Fluid Mech.* **2001,** *433,*77–104.

50. Yamamoto Y.; Potthoff, M.; Kajishima,T.;and Tsuji Y. Large-eddy simulation of turbulent gas-particle flow in a vertical channel: effect of considering inter-particle collisions. *Fluid Mech.***2001,***442,* 303–334.

CHAPTER 8

RECOMMENDATIONS FOR DESIGNING, CALCULATION, AND INDUSTRIAL USE OF A DYNAMIC SCRUBBER

CONTENTS

8.1   INTRODUCTION

Parameters of any industrial structures are defined, as it is known, first of all by a technological level of their designs. Meanwhile, objectively it is necessary to recognize that designing a gas-cleaning installation of buildings in Russia as a whole essentially loses world level and elimination of this defect occurs is inadmissible slow rates. Struggle against an atmospheric pollution and protection against aftereffects of this pollution represents a multidisciplinary problem. Organizational—technical and scientific bases of its solution are still far from perfect [1–12]. It is possible with sufficient basis to tell that creation of basic new processes with the closed dust-laden gas streams—business concerning the remote prospect. In the future, it is necessary to be oriented on such ways of protection of air basin which are realizable means of modern techniques.

Main is the decrease in volumes of the flying emissions during the basic operating procedure. Designers of gas-cleaning installation buildings should pay attention that complexity of problems solved by them is quite often aggravated with weak responsibility of the persons designing the basic manufacture (Belevitskij, 1990; Bogatich, 1978; and Rusanov, 1969). Not too seldom economic benefit attained in sphere the basic manufacture, is completely recoated by costs on clearing of huge volumes of the flying emissions. There are situations when building of gas-cleaning installation buildings in general appears almost impossible. At last, it is necessary to pay attention once again that gas-cleaning installation buildings cannot be observed as panacea, which will correct all flaws of the specialists devising production engineering of the basic manufacture. Unfortunately, such point of view existed throughout many years and as a result on a row of the factories the gas-cleaning installation buildings which still fairly have not attained term of the operational deterioration, have ceased to cope with the increased technological loading. The role and a place of gas-cleaning installation buildings in system of provisions on aerosphere protection consist in liquidating and neutralizing those emissions which formation cannot be prevented any preventive measures. Such statement of a question is dictated by elementary economic reasons. Under the world data, cost of gas-cleaning installation buildings makes from 10 to 40–45 per cent (in certain cases even to 50%) in relation to cost basic dust-laden, the equipment and, in connection with toughening of sanitary demands, tends to the further growth.

It is necessary to note defects, to correct which is necessary in the proximal years.

1. Insufficiency of the nomenclature of a gas-cleaning installation and its lag from growing powers of the industry.

2. Weakness of design baseline in which predominates empiric the analysis.

3. Absence of strict scientific criteria for designing of gas-cleaning installation buildings with number of steps of clearing two and more. For the specified reason at designing of such buildings, the big role is played by purely heuristic factor.

4. The weakest and not an authentic level of scrutiny of the questions connected with drawing by the flying emissions of a damage to a circumambient and, accordingly, with definition of economic benefit of liquidation of this damage.

These and other unresolved problems should be solved that who initiates today to master designing of gas-cleaning installation buildings.

8.2   HYDRODYNAMIC PROBLEMS OF DESIGNING A GAS-CLEANING INSTALLATION OF BUILDINGS

The gas-cleaning installation building consisting of a complex of consistently working apparatuses and communications represents the aerodynamic system possessing a row of features:

1. In most cases, through a channel of clearing of gas emissions gas, and an aerosol (firm, liquid, mixed) moves not. Industrial aerosols always polydisperse aerosol. If the hydrodynamics of dispersion streams, in general, is difficult enough, in a unequigranular stream (Idelchik, 1983), it repeatedly becomes complicated.

2. At passage through a channel of clearing of gas emissions, the aerosol stream continuously undergoes changes of speed and a traffic route, flows round obstructions, and overcomes channels of various configurations. Veering can be smooth, sudden, and on any angle. Specific types of channels are foam layers on lattices of foamy apparatuses. As special aspects of obstructions, cyclone separators where on a short piece of a way and for very small time, the stream undergoes many affecting serve.

3. In gas-cleaning installation buildings, often there are so-called *free flooded streams* (Idelchik, 1968). They originate, for example, at an entry of a stream from a gas pipeline in the apparatus of much larger cross section already filled with gas. Thus, between an inducted stream of a basin (aerosol) and a mix in the apparatus, there are difficult contacts, where the result is quite often expressed in not design fall of aerosol corpuscles. The specified phenomenon not always is desirable. For example, in the scrubber of full transpiration used as the air conditioner, but not as the dust precipitator, the bottom part turns to "the dry" dust-collecting chamber.

4. Special complexity in aerodynamics gas-cleaning installation, a building consists that in many cases, it is necessary to observe dynamics of gas and dispersion phases of an aerosol separately (Bogatich, 1978; and Dubinsky, 1977). Aerosol corpuscles, except for the smallest (2–3μm) owing to the time lag and long relaxation time have no time to follow behind gas-phase motion, and deviate it, sometimes so sharply that it is hardly probable not solving image influences parameters of dust removal apparatuses. Not taking into account the given factor leads to coarse design errors and quite often forces to bring in expensive alterations to already working buildings.

In designing of gas-cleaning installation buildings, it is necessary to analyze and solve three basic hydroaerodynamic problems: calculation of a water resistance of a channel, sampling of fan cars, and definition of places of their arrangement; provision on all channels of clearing of gas emissions of a regime of motion of the gas in the best way answering to set conditions; provision of a uniform distribution of gas and dust loading between apparatuses and in them [13–25].

8.3 FEATURES OF DESIGNING OF WET GAS-CLEANING INSTALLATIONS

The type of wet apparatuses is rather extensive and superfluous. Its only small part is issued in the form of the normalized rows, and serial exhaustion is restricted by only several types. Many apparatuses are made as the nonstandard equipment at the factories where it is possible to place the order, or is direct on an assembly site. In the latter case, the workmanship appears low. Wet apparatuses can carry out the following functions:

To chill the gas (aerosol); as alternative—with salvaging of warmth of an irrigation water which in process heats up;

To moisten to (condition)—gas (aerosol) before its supply on clearing; this process is often carried out in a regime of full transpiration of a chilling liquid;

To absorb gas or steam components from gas emission or from an aerosol disperse medium;

To trap firm and liquid corpuscles of a dispersoid of an aerosol.

These functions in many cases are carried out simultaneously though the clearing purposes cannot be provided. The account of collateral functions, including contradiction of the designing purposes, is absolutely obligatory. They can change all kinetics of process and change the properties of a circulating opening to such extent that it can demand its additional difficult machining.

In wet apparatuses, heat-and-mass transfer that occur and their completeness depend on character and intensity of contact of phases. Technological calculation of the wet apparatus in the fullest aspect develops three independent calculations: heat exchange between an irrigation water and medium in the apparatus, a mass transfer (absorption of gases and steams a liquid, liquid transpiration), and trapping of aerosol corpuscles [26–30].

The product trapped in gas-cleaning installations can be in two conditions:

In the form of a liquid—if during clearing there was only absorption of components of a gas phase of emission or if the fog, d.h. a dispersoid liquid an aerosol was trapped;

In the form of sludge—if in the wet apparatus there was a dust trapping, d.h. a dispersoid forms an aerosol;

The liquid is accepted in the factory which uses it at own discretion, or goes to manufacturing system of clearing of flows, or passes local clearing within a gas-cleaning installation building and again is fed on apparatus irrigation (the closed cycle of an irrigation).

Sludge is carried on a mud space where parches can be used, or is passed through system of settlers and fine gauge strainers; after a filtering the liquid is refunded on an irrigation, and the filtered off weight (in the form of so-called cakes) is reclaimed.

8.4   SYSTEM OF WATER RECYCLING

Process flowsheets of preparation of an opening also are so various, how much various the problems solved by wet clearing of gas emissions. The momentous factor is selectivity of an irrigation (on regimes and a chemical compound of irrigation waters). Both on separate apparatuses, and in one apparatus on different knots of an irrigation.

Irrigation of wet apparatuses without an opening reuse is applied now seldom as conducts to unduly to the big charge of a liquid and reagents containing in it. In designs cyclicity of an irrigation, multiple use of the same opening with its gradual partial deductions from a cycle and the additive of a fresh opening is usually provided. If hot gas and a circulating opening are exposed to clearing heats up, the heat exchanger—refrigerator is built in a cycle.

At designing of sprinkling systems, the momentous role is played by the concept of a *limiting condition* of an opening. If to irrigate the wet apparatus with a circulating opening without removal of its part and the additive fresh after a while, the opening condition expels its further use. The limiting condition can be defined by the following factors:

1. If the opening traps a dispersoid firm in an aerosol the suspended matter should not exceed concentration above which work of sprinklers is broken. Other criterion of a limiting condition in this case is inadmissible decrease in extent of the trapping, called by the big removal of a trapped product with splashes of the concentrated suspended matter.

2. At accumulation in an opening of some components, their crystallization on an internal surface of pipes, apparatuses, armatures is under certain conditions initiated, and it is accompanied also by sedimentation of inert suspended matters. The crystallization beginning means that there has stepped a limiting condition of an opening; its further use will lead to rapid driving down of elements of an irrigation system.

3. At absorption of steams or gases a limiting condition is such saturation of an opening at which its further use loses meaning: between an opening and an absorbed component balance is installed, and absorption stops.

It is necessary to pay attention of designers to condition which is specific to clearing of gas emissions. The removal in an aerosphere of a trapped component occurs for two reasons: first, a component quantity is not entrained by a liquid; secondly, the component entrained by a liquid partially is taken out from the apparatus with splashes.

8.5   RECOMMENDATIONS ABOUT DESIGNING

Recommendations are resulted only on packaging a twirled air swirler:
1. Diameter of the apparatus is designed proceeding from productivity on gas

$$D = 1,26\sqrt{\frac{Q}{W}}$$

2. Air swirler outside diameter is accepted equal $D` = (0.75-0.85)\,D$
3. The number of guide vanes is computed proceeding from diameter of an air swirler $z = (10 + 25)\,D`$ and rounded off to the number convenient for staggered pitch length of a round on equal parts.
4. The length of blades is designed on a relationship

$$l_\Lambda = \frac{0,5D_i \sin\dfrac{360}{z_i}}{\sin(\alpha_y + \dfrac{360}{z_i})}.$$

And the angle of installation of blades αy is recommended to be accepted $35 + 45°$.

5. For the set size of corpuscles, it is defined critical angular speed of twirl of an air swirler

$$\omega_{opt} = 397.38 \cdot w^{1.65}(m \cdot 10^6)^{0.31} \bar{D}^{-0.31} \cdot \bar{z}^{1.05} \exp\left[-0.018 \cdot 10^6 d_p - (1,06 + 0,034w) \cdot \cos\alpha - 2,18 \cdot \cos^2\alpha\right]$$

6. The air swirler direction of rotation is recommended such, at which it is 90°.
7. The air swirler water resistance is defined by the formula

$$\Delta\rho_{\omega>0} = \frac{0,5\rho\omega D_1(0,5\omega D_1 - W_1 \cos\alpha)}{1 + \dfrac{1,5 + 1,1\alpha/90°}{z(1 - d_1^2)}}$$

8. Power of the drive is roughly sized up on dependence:
$$P = Q \cdot \Delta\rho$$

8.6   CLEARING OF GASES OF DUST IN THE INDUSTRY

The had results hardware in manufacture of roasting of limestone at conducting of redesign of system of an aspiration of smoke gases of baking ovens. The devised scrubber is applied to clearing of smoke gases of baking ovens of limestone in the capacity of a closing stage of clearing.

Temperature of gases of baking ovens in main flue gas breeching before a copper utilizatorom 500–600°C, after exhaust-heat boiler 250°C. An average chemical compound of smoke gases (by volume): 17 per centCO_2; 16 per centN_2; 67 per cent CO. Besides, in gas contains to 70 mg/m³ SO_2; 30 mg/m³ H_2S; 200 mg/m³F, and 20 mg/m³ CI. The gas dustiness on an exit from the converter reaches to 200/m³ the Dust, as well as at a fume extraction with carbonic oxide after-burning, consists of the same components, but has different maintenance of oxides of iron. In it than 1micron, than in the dusty gas formed at after-burning of carbonic oxide contains less corpuscles a size less. It is possible to explain it to that at after-burning CO raises temperatures of gas and there is an additional excess in steam of oxides. Carbonic oxide before a gas heading on clearing burns in the special chamber. The dustiness of the cleared blast-furnace gas should be no more than 4 mg/m³.

The following circuit design (Figure 8.1) is applied to clearing of the blast furnace gas of a dust.

Gas from a furnace mouth of a baking oven 1 on gas pipes 3 and 4 is taken away in the gas-cleaning plant. In raiser and downtaking duct gas is chilled, and the largest corpuscles of a dust which in the form of sludge are trapped in the inertia sludge remover are inferred from it. In a centrifugal scrubber 5 blast furnace gas is cleared of a coarse dust to final dust content 5–10/m³ the dust drained from the deduster loading pocket periodically from a feeding system of water or steam for dust moistening. The final cleaning of the blast furnace gas is carried out in a dynamic spray scrubber where there is an integration of a finely divided dust. Most the coarse dust and drops of liquid are inferred from gas in the inertia mist eliminator. The cleared gas is taken away in a collecting channel of pure gas 9,whence is fed in an aerosphere. The clarified sludge from a gravitation filter is fed again on irrigation of apparatuses. The closed cycle of supply of an irrigation water to what in the capacity of irrigations the lime milk close on the physical and chemical properties to composition of dusty gas is applied is implemented. As a result of implementation of trial installation clearings of gas emissions the maximum dustiness of the gases which are thrown out in an aerosphere, has decreased with 3,950mg/m³to 840mg/m³, and total emissions of a dust from sources of limy manufacture were scaled down about 4,800to/a to 1,300to/a.

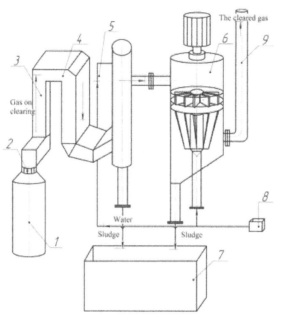

FIGURE 8.1   Process flowsheet of clearing of gas emissions: (1)—bake roasting; a (2)—water block; a (3)—raiser; (4)—downtaking duct; a (5)—centrifugal scrubber; a (6)—scrubber dynamic; a (7)—forecastle of gathering of sludge; a (8)—hydraulic hitch; and a (9)—chimney.

Such method gives the chance to make gas clearing in much smaller quantity, demands smaller capital and operational expenses, and reduces an atmospheric pollution and allows to use water recycling system.

8.7 CONCLUSIONS

1. Modern problems of designing of gas-cleaning installation buildings are observed. Specific features of process of clearing of gas emissions are in detail presented.

2. The solution of an actual problem on perfection of complex system of clearing of gas emissions and working out of measures on decrease in a dustiness of air medium of the industrial factories for the purpose of betterment of hygienic and sanitary conditions of work and decrease in negative affecting of dust emissions on a circumambient is in-process given.

3. Designs on modernization of system of an aspiration of smoke gases of baking ovens of limestone with use of the new scrubber which novelty is confirmed with the patent for the invention are devised. Efficiency of clearing of gas emissions is raised. Power inputs of spent processes of clearing of gas emissions and power savings at the expense of modernization of a flowchart of installation of clearing of gas emissions are lowered.

KEYWORDS

- **Smoke gases**
- **The baking oven**
- **The closed cycle**
- **The industry**
- **The plant flow diagram**
- **Water recycling**

REFERENCES

1. Belevitskij, A.M.; Designing a Gas-Cleaning Installation of Constructions. L: Chemistry;**1990.**

2. Berezhinsky, A. I.;and Homutinnikov, I. C.; Salvaging, Cooling and Clearing of Smoke Gases. M.: Metallurgy;**1967.**

3. Bogatich, C. A.; Complex Machining of Air in Foamy Apparatuses. L: Shipbuilding;**1978.**

4. Bogatich, C. A.; Cyclonic and Foamy Apparatuses. L: Engineering Industry;**1978.**

5. Valdberg, A.J.; Isjanov, L.M.;and Tarat, E.J.; Dust Separation Production Engineering.L: Engineering Industry;**1985.**

6. Effect of air pollutions on vegetation.In: The Reasons, Affectings, Retaliatory Measures. Ed. Dessler, X.;M.: The Forest Industry;**1971.**

7. The Gas-Cleaning Installation Equipment. Bag Hoses: Catalogue M.;**1985.**

8. Guderian, R.; Pollution of Air Medium: Transfer with English.M.: The World;**1979.**

9. Dubtchik, R. V.; Rehash of a Waste of Aluminium Manufacture Abroad. M.;**1978.**

10. Dubinsky, F. E.;and Lebedjuk, G.K.; Scrubbers of the Venturi. Sampling, Calculation, Application. M.: Chemical and Oil Engineering Industry;**1977.**

11. Dubinsky, F. E.; et al. Clearing of Gases of Iron Cupola Furnaces. AChemical and Oil Engineering IndustryM.;**1978.**

12. Idelchik, I.E.; Fluid Kinetics of Industrial Apparatuses. M.: Engineering Industry;**1983.**

13. Idelchik, I.E.; Alexanders, V. P.;and Kogan, E.I.; Research of direct-flow cyclone separators of system of an ash collection of a state district power station.*Heat Power Eng.***1968,***8,* 45–48,*(in Russian).*

14. Intensive Columned Apparatuses for Machining of Gases by Liquids. Ed. Tarata, E. J.; L: Publishing House i Lie;**1976.**

15. Lebeduk, K. E.; et al. Methods of Clearing of Kiln Gases from a Dust. Chemical and Oil Engineering IndustryM.;**1971.**

16. Pazin, L. M.;and Libina, V.L.; Industrial and Sanitary Clearing of Gases. AChemical and Oil Engineering Industry. M.;**1977,***5,* 2–3, *(in Russian).*

17. Potapov, I.P.;and Kropp, L.D.; Batarejnye Cyclone Separators.M.: Energy;**1977.**

18. Rakhmonov, T. Z.; Salimov, S.C.;and Umirov, R. R.; Wet Clearing of Gases in Apparatuses with a Mobile Nozzle.T: The Fan;**2005.**

19. Rusanov, A.A.; Urbah, I.T.;and Anastasiadi, A.P.; Clearing of Smoke Gases in Industrial Power Engineering. M.: Energy;**1969.**

20. Sagin, B.S.;and Gudim, L.I.; Dedusters with the Counter Twirled Streams. The Chemical Industry.Moscow;**1984,** *8,* 50–54,*(inRussian).*

21. The Directory after a Heat—and to an Ash Collection. Rusanov's, A. A.;M.; Energy;**1975.**

22. Stark, S.B.; Dust Separation and Clearing of Gases in Metallurgy. M.: Metallurgy;**1977.**

23. Staritsky, V. A.; The Gas Equipment of Factories of Ferrous Metallurgy. M.: Metallurgy;**1973.**

24. Ugov, V.N.;and Valdberg, A.J.; Clearing of Gases Wet Fine Gauge Strainers. M.: Chemistry;**1972.**

25. Ugov, V.N.; Clearing of Industrial Gases by Electrostatic Precipitators. M.: Chemistry;**1967.**

26. Ugov, V.N.;and Valdberg, A.J.; Preparation of Industrial Gases for Clearing. M.: Chemistry;**1975.**

27. Ugov, V.N.;and Myagkov, B.I.; Clearing of Industrial Gases by Fine Gauge Strainers. M.: Chemistry;**1970.**

28. Ugov, V.N.; Valdberg, A.J.; Myagkov, B.I.;and Rashidov, I.K.; Clearing of Industrial Gases of a Dust. M.: Chemistry;**1981.**

29. Directions and norms of technological designing and technical-and-economic indexes of a power equipment of the factories of ferrous metallurgy. Metal works. T. **18**. It is sewn up aerospheres. Clearing of gases of a dust. Sanitary codeno 1-41-00. SSSR. **2001.**

30. Directions for to Dispersion Calculation in an Aerosphere of the Harmful Substances Containing in Emissions of the Factory. Sanitary code no 369-04. M.: Building;**2005.**
31. Economy of Sterilisation of Gas Emissions. Chemical and Oil Engineering IndustryM.;**1979,**$6,$25.

PART II

DUST EXTRACTORS OF SHOCK-INERTIAL ACT

CHAPTER 9

ORGANIZATION OF HYDRODYNAMIC INTERACTION OF PHASES IN DUST EXTRACTORS WITH INNER CIRCULATION OF A LIQUID

CONTENTS

9.1   INTRODUCTION

The main problems of the known wet-type collectors are single-value usage of liquid in dust removal process and its large charges for gas clearing. For machining of great volumes of an irrigating liquid and slimes, salvaging bulky complex systems of circulating water supply is required. It considerably increases the cost of process of clearing of gas and does its commensurable with clearing the cost at application of electrofilters and bag hoses (Rodions, 1985; and Nuкiyama, 1938).

Therefore, now there is a necessity for creation of such wet-type collectors which would work with the low consumption of an irrigating liquid. New dedusters should combine the basic advantages of modern means of clearing of gases: simplicity and compactness, high efficiency, possibility of management of processes of a dust separation, and optimization of regimes [1–10].

To use modern techniques in gas industry, wide circulation of liquid is used in wet-type collectors with inner circulation in systems of gas cleaning in Russia and abroad (Ugov, 1981; and Shcwidki, 2002).

9.2   SURVEY OF KNOWN BUILDS OF SCRUBBERS WITH INTERNAL CIRCULATION OF A LIQUID

System will improve considerably if water circulation is performed. Accumulated slurry can be drained continuously or periodically or by means of mechanical carriers, in this case necessity for water recycling system disappears, or a hydraulic path—a drain of a part of water. In the latter case, the device of system of water recycling can appear expedient, but load on it is much less, than at circulation of all volume of water (Hertzian, 1978; and Hertzian, 1975).

Dust traps of such aspect are characterized by the presence of the capacity filled with water. Cleared air contacts to this water and contact conditions are determined by interaction of currents of air and waters. The same interaction calls water circulation through a zone of a contact at the expense of energy of the most cleared air [11–14].

The water discharge is determined by its losses on transpiration and with deleted slurry. In slurry removal by mechanical scraper carriers or manually the water discharge minimum also makes only 2–5g on $1m^3$ air.

At periodic drain of the condensed slurry, the water discharge is determined by consistency of slurry and averages to 10 r on 1 m³air, and at fixed drain the charge does not exceed 100–200 g on 1 m³air. Filling of dust traps with water should be controlled automatically. Maintenance of a fixed level of water has primary value as its oscillations involve essential change as efficiency and productivity of system [15–17].

The basic most known constructions of these apparatuses are introduced in Figure9.1 (Kousov, 1993).

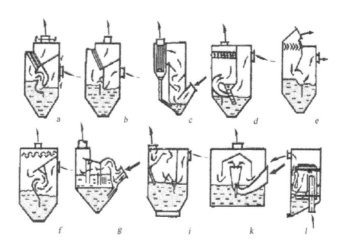

FIGURE 9.1   Constructions of scrubbers with inner circulation of a fluid: (a)—rotoklon N (USA); (b)—PVM CNII (Russia); (c)—a scrubber a VNIIMT (Russia); (d)—a dust trap to me (Czechoslovakia); (e)—dust trap WNA (Germany); (f)—dust trap "Asco" (Germany); (g)—dust trap LGP (Russia); (i)—dust trap "Klayrator" (USA); (k)—dust trap VDN (Austria); (l)—rotoklonRPA a NIIOGAS (Russia).

Mechanically each of such apparatuses consists of contact channel fractionally entrained in a fluid and the drip pan merged in one body. The principle of performance of the apparatuses is based on intensive wash down of gases in contact channels of various configurations with the subsequent separation of a water gas flow in the drip pan. The used fluid is not discharged and recirculates several times for dust removal process.

Circulation of a liquid in the wet-type collector occurs at the expense of a kinetic energy of a gas stream.

Each apparatus is equipped with some devices for fixing of fluid level and for removal of slurry from the scrubber collecting hopper.

Distinctive features of apparatuses:

1. Fluid spray in the gas without the use of injectors allows to use a fluid with the high contents of suspended matters for spraying (up to 250mg/m³).

2. Landlocked fluid circulation in apparatuses allows reusing a fluid in contact devices of scrubbers and by that to device out its charge on clearing of gas to 0.5kg/m³, that is, in 10 and more times in comparison with other types of wet-type collectors.

3. Removal of the trapped dust from apparatuses in the form of slurries with low humidity that allows to simplify dust salvaging to reduce loading by water purification systems. In certain cases, it is possible to refuse a construction of system of water purification.

4. Layout of the drip pan in a body of the apparatus which allows diminishing the sizes of dust traps to supply their compactness.

The indicated features and advantages of such scrubbers have led to wide popularity of these apparatuses, active working out of various constructions, research, and a heading of wet-type collectors, as in Russia and abroad.

The scrubbers are presented in Figure9.1. Concern to apparatuses with noncontrollable operating conditions as in them there are no gears of regulating. In this type of scrubbers, the stable conditions of activity of a high performance are difficultly supplied, especially at varying parameters of cleared gas (pressure, temperature, a volume, a dust content, etc.). In this regard, wet scrubbers with controlled variables are safer and better in maintenance. Regulation of operating conditions allows changing hydraulic resistance from which magnitude, according to the power theory of a wet dust separation, efficiency of trapping of a dust depends. Regulation of parameters allows to maintain dedusters in an optimum regime. Optimum conditions of interaction of phases are thus provided and peak efficiency of trapping of a dust with the least power expenses is reached. The great value is acquired by dust traps with adjustable resistance also for stabilization of processes of gas cleaning at varying parameters of cleared gas. A set of these scrubbers are shown in Figure9.2.

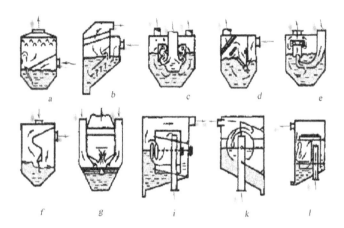

FIGURE 9.2   Apparatuses with controlled variables: (a)—under the patent no 1546651 (Germany), (b)—the ACE no 556824 (USSR), (c)—the ACE no 598625 (USSR), (d)—the ACE no 573175 (USSR), (e)—under the patent no 1903985 (Germany), (f)—the ACE no 13686450 (France), (g)—the ACE no 332845 (USSR), (i)—the ACE no 318402 (USSR), (k)—the ACE no 385598 (USSR), and (l)—type RPA a NIIOGAS (USSR).

In the scrubbers presented in Figure9.2, and, turn of controlling partitions is made either manually, or is distant with the drive from the electric motor, and in a dust trap on Figure9.2, in manually, moving of partitions rather each other on a threaded connection. In dust traps in Figure9.2(e). The lower partitions are mounted on a floating structure able to stabilize clearing process at different levels of the fluid.

The applied principle in dust traps is presented in Figures 9.2 and 9.1. Contact devices in these apparatuses are had on a midwall of the floating chamber entrained in a liquid. Contact devices are fixed in a casing of the apparatus by means of hinges. Such construction of dust traps support automatically to constants of an apparatus hydraulic resistance at varying gas load.

Well known internationally (in the USA), the bubbling deduster that has been prepackaged in one block with the centrifugal fan (Shcwidki, 2002). Such apparatus is named type N rotoklon (see Figure 9.3).

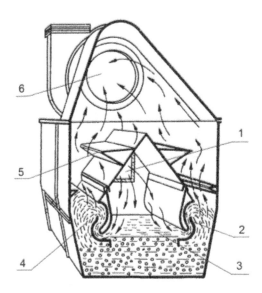

FIGURE 9.3   Type N rotoklon: (1)—the device for gaseous feed; (2)—guide vanes;(3)—a water level;(4)—a cleaning tank; (5)—the drip pan; and (6)—the device for gas deductions.

Gas passes through the slot-hole channels (impellers) formed by bent shovels. The bottom part of blades is hauled down in a liquid. The curtain from drops of a spluttered liquid is thus created. In this curtain, gas is chilled and cleared of a dust. Speed of gas in slot-hole channels of a rotoklon does not exceed 15m/s. The liquid in the apparatus is supported at certain level which plays the momentous role in normal maintenance of a rotoklon. The type rotoklon is intended for clearing of outtake air of fibrous and sticky dust. Sludge is drained from the apparatus periodically in process of accumulation in it of certain quantity of a dust. For compensation of ablation and transpiration of water it feed in a rotoklon in the quantity which is not exceeding 0.03 kg/m³. Productivity of rotoklons from 2.5 to 90 thousand m³/ch. Gas loading accepts proceeding from the square of a mirror of water in the basin. It is equal on the average 1,000 m³/(ch·m²). The length of the slot-hole channel makes from 0.8 to 15 m. The water resistance of a rotoklon does not exceed 1.5–2.0 kpa(Kucheruk, 1963; and Straus, 1981).

In Russia, following bubbling dedusters are applied: the institute builds, working as rotoklon*N*, productivity 10 and 40 thousand m³/ch; builds ITS of type AWC productivity 45 both 10 thousand m³/ch and builds of institute ITS "industrial acceptance" of type AWC (Figure 9.4 see) productivity 3,5; 10; 20 and 40 thousand m³/ch. Apply also a rotoklon of type "Ural" at which gas loading on 1 m of the slot makes 10–15 thousand m³/chfor a water resistance 1,000 Pa(Shcwidki, 2002).

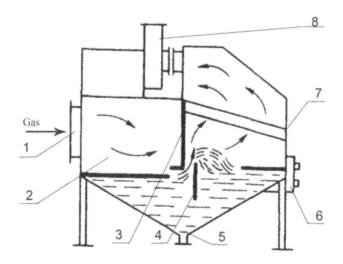

FIGURE 9.4   Deduster of type AWC:(1)—the upstream end; (2)—the body; (3)—the overhead partition; (4)—the bottom partition; (5)—a drain connecting pipe; (6)—the device for water level maintenance; (7)—the drip pan; and (8)—the ventilating fan.

From the literary data, it is known that scrubbers with internal circulation of a liquid can work in a narrow interval of change of speed of gas in contact channels. Scrubbers are used in industrial production for clearing of gases of a dust in systems of an aspiration of auxiliaries (Hertzian, 1978; Kousov, 1993; and Kucheruk, 1963). Known apparatuses are rather sensitive to change of gas load on the contact channel and to fluid level, negligible aberrations of these parameters from best values lead to a swing of levels of a fluid at contact channels, to unstable operational mode and dust clearing efficiency lowering. Because of low speeds of gas in contact channels known apparatuses have large gabarits. These defects, and also weak level of scrutiny of processes passing in apparatuses, absence of reliable methods of their calculation complicates working out of new rational designs of wet-type collectors of the given type and their wide introduction in manufacture. In this regard, necessity of more detailed theoretical and experimental studies of scrubbers with inner circulation of a fluid for the purpose of the prompt use of the most effective and cost-effective constructions in systems of clearing of industrial gases has matured.

9.3 THE ORGANIZATION OF HYDRODYNAMIC INTERACTING OF PHASES

In scrubbers with inner fluid circulation, the process of interaction of gas, liquid, and solid phases consist of solid phase (dust), finely divided in gas, passes in a fluid implements. Because concentration of a solid phase in gas has rather low magnitudes (to 50 g/m^3), it does not influence the hydrodynamics of streams (Nuкiyama, 1938). Thus, hydrodynamics study in a scrubber with inner circulation of a fluid has paid less attention to gas- and liquid-phase interactions.

Process of hydrodynamic interaction of phases can be observed as stages passing consistently:
- Fluid acquisition by gas flow on the contact device influent
- Fluid distribution by a fast-track gas flow in the .contact channel
- Integration of fluid drops on the contact device effluent
- Separation of a liquid from gas in the time of passage through a trap of drops.

9.3.1 FLUID ACQUISITION BY A GAS FLOW ON THE CONTACT DEVICE INFLUENT

Before an entry in the contact partition of the apparatus there is a contraction of a gas flow for gas speed enhancement, acquisition of high layers of a fluid and its entrainment in the contact channel. Functionability of all dust trap depends on efficiency of acquisition of a fluid by a gas flow—without fluid acquisition will not be supplied effective interaction of phases in the contact channel and, hence, qualitative clearing of gas of a dust will not be attained. Thus, fluid acquisition by a gas flow on an entry in the contact device is one of the defined stages of hydrodynamic process in a scrubber with inner circulation of a fluid. Fluid acquisition by a gas flow can explain the presence of interphase turbulence which is advanced on an interface of gas and liquid phases. Conditions for origination of interphase turbulence are the presence of a gradient of speeds of phases on boundaries, difference of viscosity of flows, and an interphase surface tension. At gas driving over a surface of a fluid, the last will be brake gas boundary layers therefore in them there are the turbulent shearing stresses promoting cross-section transfer of energy. Originating cross-section turbulent oscillations lead to penetration of turbulent gas curls into boundary layers of a fluid with the subsequent illuviation of these stratums in curls. Mutual penetration of curls of boundary layers leads as though to the clutch of gas with a fluid on a phase boundary and to entrainment of high layers of a fluid for moving gas over its surface. Intensity of such entrainment depends on the kinetic energy of gas flow, from its speed over a fluid at an entry in the contact device. On gradual increase in speed of gas, there is a change of a surface of a fluid at first from smooth to undular, then ripples are organized and, at last, there is a fluid dispersion in gas. The efficiency of wet-type collectors with inner fluid circulation is expedient for conducting by means of a parameter $m = V_z/V_g$ m³/m³ equal to the ratio of volumes of liquid and gas phases in contact channels and characterizing the specific charge of a fluid on gas irrigating in channels. Obviously that magnitude m will be determined, first of all, by speed of a gas flow on an entry in the contact channel. Other diagnostic variable is fluid level on the contact channel influent which can change cross section of the channel and influence speed of gas.

Thus, for the exposition of acquisition of a fluid a gas flow in contact channels it is enough to gain experimental relation of following type:

$$m = f(W_r, h_g) \qquad (9.1)$$

9.3.2 FLUID SUBDIVISION BY A FAST-TRACK GAS FLOW IN THE CONTACT CHANNEL

As shown further, efficiency of trapping of a dust particle in many respects depends on the size of fluid drops: The decrease of drop size will lead to increase in the dust clearing efficiency. Thus, the given stage of hydrodynamic interacting of phases is rather important.

Process of subdivision of a fluid by a gas flow in the contact channel of a dust trap occurs at the expense of high relative speeds between a fluid and a gas flow. For calculation of average diameter of the drops gained in contact channels, it is expedient to use the empirical formula of the Japanese engineers Nukiymas and Tanasavas who consider agency of operating conditions along with physical performances of phases:

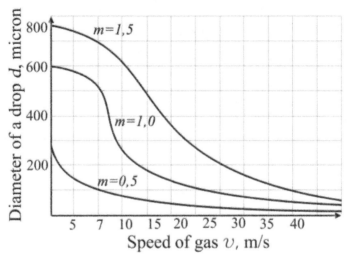

FIGURE 9.5 Relation of an average size of drops of water in blade impellers from speed of gas.

$$D_o = \frac{585 \cdot 10^3 \sqrt{\sigma}}{W_r} + 49,7 \left(\frac{\mu_l}{\sqrt{\rho_l \sigma_l}} \right)^{0,2} \frac{L_l}{V_r} \qquad (9.2)$$

where W_r is the relative speed of gases in the channel, mps; σ_l is the factor of a surface tension of a fluid, N/m; ρ is the fluid density, kg/m^3; μ_l is the viscosity of a fluid, Pascal/s; L_l is the volume flow of a fluid, m^3/ s; V_r is the volume flow of gas, m^3/s.

On Figure9.5, calculated curves of average diameter of drops of water in contact channels depending on speed of a gas flow are resulted. Calculation is conducted by formula (9.2) at following values of parameters: σ= 720·10^3 N/m; ρ= 1,000 kg/m^3; μ = 1,01 … 10^2 Pa/s.

The gained relations prove that the major operating conditions with average size of drops are speed of gas flow W_r and the specific charge of a fluid on gas irrigating m. These parameters determine hydrodynamic structure of an organized water gas flow.

With increasing speed of gas process of subdivision of a fluid by a gas flow gains in strength, and drops of smaller diameter are organized. The most intensive agency on a size of drops renders change of speed of gas in the range from 7 to 20 mps, at the further increase in speed of gas (> 20 mps) intensity of subdivision of drops is reduced. It is necessary to note that in the most widespread constructions of shock-inertial apparatuses (rotoklons N) which work at speed of gas in contact devices of 15mps, the size of drops in the channel is significant and makes 325–425μm. At these operating conditions and sizes of drops, qualitative clearing of gas of a mesh dispersivity dust is not attained. For decreasing particle size and increasing efficiency of these apparatuses the increase in speed of gas to 30, 40, and 50 mps and more depending on type of a trapped dust is necessary.

The increase in the specific charge of a fluid at gas irrigating leads to growth of diameter of organized drops. So, at increase m with $0.1 \cdot 10^3$–3 ... 10 m^3/m^3 the average size of drops is increased approximately at 150μm. For security of minimum diameter of drops in contact channels of shock-inertial apparatuses, the specific charge of a fluid on gas irrigating should be optimized in the interval $(0.1–1.5) \cdot 10^3$ m^3/m^3. It is necessary to note that in the given range of specific charges with a high performance the majority of fast-track wet-type collectors works.

9.3.3   INTEGRATION OF FLUID DROPS ON THE CONTACT DEVICE EFFLUENT

On an exit from the contact device, there is an expansion of the irrigated gas and increase in drops of a liquid at the expense of their concretion. The maximum size of the drops weighed in a gas flow is determined by stability conditions: the size of drops will be that more than less speed of a gas flow. Thus, on an exit from the contact device effluent together with fall of speed of a gas flow the increase in a size of drops will be observed. Turbulence in a dilated part of a stream more than in the channel with constant cross section. Turbulence grows with increase in an angle of expansion of a stream, and it means that speed of turbulent concretion will grow also with increase in an angle of expansion of a stream. Than more intensity of concretion of corpuscles of a liquid, the corpuscle on an exit from the contact device will be larger and the more effectively they will be separated in the drip pan (Hertzian, 1975).

Practice shows that the size a coagulation of drops on an exit makes of the contact device, as a rule, more than 150μm. Corpuscles of such size are easily trapped in the elementary devices (the inertia, gravitational, centrifugal, etc.).

9.3.4   BRANCH OF DROPS OF A FLUID FROM A GAS FLOW

The inertia and centrifugal drip pans are applied to branch of drops of a fluid from gas in shock-inertial apparatuses in the core. In the inertia drip pans, the branch implements at the expense of veering of a water gas flow. Liquid drops, moving in a gas flow, possess definitely a kinetic energy; thanks to which at veering of a gas stream they by inertia move rectilinearly and are inferred from a flow. If to accept that the drop is in the form of a sphere and speed of its driving is equal in a gas flow to speed of this flow the kinetic energy of a drop, moving in a flow, can be determined by formula

$$E_{\hat{e}} = \frac{\pi D_0^{\;3}}{6} \rho_l \frac{W^2_{\;r}}{2} \qquad (9.3)$$

with decrease of diameter of a drop and speed of a gas flow the drop kinetic energy is sharply diminished. At gas-flow deflection the inertial force forces to move a drop in a former direction. The more the drop kinetic energy, the more is the inertial force:

$$E_{\hat{e}} = \frac{\pi D_0^{\;3}}{6} \rho_l \frac{dW_r}{d\tau} \qquad (9.4)$$

Thus, with flow velocity decrease in the inertia drip pan and diameter of a drop the drop kinetic energy is diminished, and efficiency drop spreads is reduced. However, the increase in speed of a gas flow cannot be boundless as in a certain velocity band of gases there is a sharp lowering of efficiency drop spreads owing to origination of secondary ablation the fluids trapped drops. For calculation of a breakdown speed of gases in the inertia drip pans it is possible to use the formula, m/s:

$$W_c = K \sqrt{\frac{\rho_{\hat{e}} - \rho}{\rho_r}} \qquad (9.5)$$

where W_c is the optimum speed of gases in free cross section of the drip pan, mps; K is the factor defined experimentally for each aspect of the drip pan.

Values of factor normally fluctuate over the range 0.1–0.3. Optimum speed makes from 3 to 5 mps (Hertzian, 1978; and Nukiyama, 1983).

9.4 PURPOSE AND RESEARCH PROBLEMS

- Creation of a dust trap with a broad band of change of operating conditions and a wide area of application, including for clearing of gases of the basic industrial assemblies of a mesh dispersivity dust.
- Creation of the apparatus with the operated hydrodynamics, allowing to optimize process of clearing of gases taking into account performances of trapped ingredients.
- Making the analysis of hydraulic losses in blade impellers and to state a comparative estimation of various constructions of contact channels of an impeller by efficiency of security by them of hydrodynamic interacting of phases;
- Determining relation of efficiency of trapping of corpuscles of a dust in a rotoklon from performance of a trapped dust and operating conditions major of which is speed of a gas flow in blade impellers. To develop a method of calculation of a dust clearing efficiency in scrubbers with inner circulation of a fluid.

KEYWORDS

- **Clearing of gas emissions**
- **Contact channels**
- **Dust**
- **Irrigation**
- **Rotoklon**
- **Separation**

REFERENCES

1. Hertzian. M. И.; Kirsanov, N.S.;and Gordon, G.M.;Teardrop and a Carryover of Liquid in a Percussion Scrubber. Works of Institute of Non-Ferrous Metals;**1978**,*44*, 71–77.
2. Hertzian, M. И.; Kirsanov, H.C.;and Gordon, G.M.; Effect of Separate Factors on Efficiency of Tapping of a Dust in Percussion Scrubbers .Works of Institute of Non-Ferrous Metals; **1975**,*36*, 112–117.
3. Kousov, P A.; Maligin, A.D.;and Skryabin, G.M.; Clearing of Gases and Air of a Dust in the Chemical Industry. St.Petersburg: Chemistry;**1993**.
4. Kusenkov, B. A.; Research of Wet-Type Collectors with Internal Circulation of Water. Water Supply and Sanitary Technics;**1971**, *11*, 19–23.
5. Kucheruk, V.V.; Clearing of Outtake Air of a Dust. Moscow;**1963**.
6. Pirumov, A.I.; Air Dust Removal. Moscow: Building Publishing House;**1974**.
7. Rodions, А.И.; Smiths, Ю.П.;and Zenkov, V.V.; Technical Equipment, Constructions, Bases of Designing of Himiko-Technological Processes of Protection of Biosphere from Plant Emissions. Moscow: Chemistry;**1985**.
8. Richkov, V.P.;and Savelyev, Ю.A.;Application of a Rotoklon "Ural"in the Industry Safety of Work in the Industry. **1982**,*8*, 43–45.
9. Straus, V.;Industrial Clearing of Gases. Moscow: Chemistry;**1981**.
10. The Directory after a Heat—and to an Ash Collection. Ed. Rusanov, A.A.; Moscow: Publishing House "Energy"; **1983**.
11. Valdberg, A.J.;and Lebedjuk, G. K.; Wet-Type Collectors of Shock and Centrifugal Moscow: Petrochemistry;**1981**.
12. Ugov, V. H.; Valdberg, A. J.;and Myagkov, B.I.;Clearing of Industrial Gases of a Dust. Moscow: Chemistry;**1981**.
13. Nuкiyama, S.;and Tanasawa, J.; Experiments on the atomisation of liquids in an air stream. *Rep. 2 Trans. Trans. Soc. Mech.Eng. (Japan).***1938**,*4(15)*.
14. Nukiyama, S.;and Tanasawa, J.; Experiments on the atomisation of liquids in an air stream. *Rep. 4 Trans. Trans. Soc. Mech.Eng. (Japan).***1938**,*5(18)*.
15. Shcwidki, V.S.;and Ladygichev, M.G.;Clearing of Gases. The Directory. Moscow: Heat Power Engineering;**2002**.
16. Shctokman, E.A.; Air Purification.Moscow: Publishing House ASV;**1998**.
17. Yudashkin, M.J.; Dust Separation and Clearing of Gases in Ferrous Metallurgy. Moscow: "Metallurgy";1984.

CHAPTER 10

EXPERIMENTAL RESEARCH AND CALCULATION OF EFFICIENCY OF SEDIMENTATION OF DISPERSION PARTICLES IN A ROTOKLON

CONTENTS

10.1 INTRODUCTION

The comparative analysis of the basic known gas-cleaning installations of impact-sluggish act shows that many builds work in a narrow range of change of speed of gas in contact channels and are used in industrial production predominantly for clearing of gases of a disperse dust in systems of an aspiration of auxiliaries. Known apparatuses are rather sensitive to change of gas loading on the contact channel and to liquid level, insignificant output disturbance of these parameters from optimum values lead to a swing of levels of a liquid at contact channels, to unstabilize operating mode and decrease in efficiency of a dust separation. Because of low speeds of gas in contact channels such devices have the big gabarits (Uzhov, 1981; and Valdberg, 1981). These deficiencies, and also weak level of scrutiny of processes proceeding in apparatuses, absence of reliable methods of their calculation complicate working out of new rational builds of wet-type collectors of the given type and their wide implementation in manufacture. In this connection has ripened necessity of more detailed theoretical and experimental studying of scrubbers of impact-sluggish act for the purpose of use of the most effective and economic builds in systems of clearing of industrial gases.

The main problems of the known wet-type collectors are single-values usage of liquid in dust removal process and its large charges for gas clearing. For machining of great volumes of an irrigating liquid and slimes, salvaging bulky complex systems of circulating water supply is required. It considerably increases the cost of process of clearing of gas and does its commensurable with clearing cost at application of electrofilters and bag hoses.

Therefore, now there is a necessity for creation of such wet-type collectors which would work with the low consumption of an irrigating liquid. New dedusters should combine the basic advantages of modern means of clearing of gases: simplicity and compactness, high efficiency, possibility of management of processes of a dust separation, and optimization of regimes.

To use modern techniques in gas industry, wide circulation of liquid is used in wet-type collectors with inner circulation in systems of gas cleaning in Russia and abroad.

10.2   EXPERIMENTAL INSTALLATION AND THE TECHNIQUE OF REALIZATION OF EXPERIMENT

The rotoklon represents the basin with water on which surface on a connecting pipe of feeding into of dusty gas the dust-laden gas mix arrives. Thus, gas changes a traffic route. The dust containing in gas penetrates into a liquid under the influence of an inertial force. Turn of blades of an impeller is made manually, rather each other on a threaded connection by means of handwheels. The slope of blades was installed in the interval 25–45 to an axis.

In a rotoklon three pairs, the blades having a profile of a sinusoid are installed. Blades can be controlled for installation of their position. Depending on the cleanliness level of an airborne dust flow, the lower lobes by means of handwheels are installed on an angle defined by operational mode of the device. The rotoklon is characterized by the presence of three channels, formation of the overhead and bottom blades. And in each following on a run of gas the channel of the bottom blade is installed above the previous. Such arrangement promotes a gradual entry of a water gas flow in slotted channels and thereby reduces the device hydraulic resistance. The arrangement of an input part of lobes on an axis with a capability of their turn allows creating a diffusion reacting region. Sequentially slotted channels mounted in a diffusion zone equipped with a rotation angle lobes, a hydrodynamic zone of intensive wetting of corpuscles of a dust. In process of flow moving through the fluid-flow curtain, the capability of multiple stay of corpuscles of a dust in hydrodynamically reacting region is supplied that considerably raises a dust clearing efficiency and ensures functioning of the device in broad bands of cleanliness level of a gas flow.

The construction of a rotoklon with adjustable sinusoidal lobes is developed and protected by the patent of the Russian Federation, which is capable to solve a problem of effective separation of a dust from a gas flow (Usmanova, 2008). Thus, water admission to contact zones implements as a result of its circulation in the apparatus.

The rotoklon with the adjustable sinusoidal lobes, presented in Figure 10.1 consists of a body (3) with connecting influent (7) and effluent (5) pipes. Moving of the overhead lobes (2) can be done by screw jacks (6), and the lower lobes (1) are fixed on an axis 8 with rotation capability. The rotation angle of the lower lobes is chosen from a condition of a persistence of speeds of an airborne dust flow. For rotation angle, regulation of a handwheel at the output parts of the lower lobes 1 is embedded. Quantity of lobe pair is determined by productivity of the device and cleanliness level of an airborne dust flow, that is, a regime of a stable running of the device. In the lower part of a body, there is a connecting pipe for a drain of slime water 9. Before connecting a pipe for a gas make 5 the labyrinth drip pan 4 is installed. The rotoklon works as follows. Depending on the cleanliness level of an airborne dust flow, the overhead lobes 5 by means of screw jacks 6, and the lower lobes 1 by means of handwheels are installed on an angle defined by operational mode of the device. Dusty gas arrives in the upstream end 7 in the upper part of the body 3 apparatuses. Having reached a liquid surface, gas changes the direction and moves to the slot-hole channel formed upper 2 and inferior 1 blades. Thanks to the traffic high speed, gas captures the upper layer of a liquid and atomizes it in small-sized drip and foam. After passage of all slot-hole channels, gas moves to the labyrinth drip pan 4 and is inferred in an atmosphere through the discharge connection 5. The collected dust settles in the loading pocket of a rotoklon and through a connecting pipe for removal of slurry 9, together with a liquid, is periodically inferred from the apparatus.

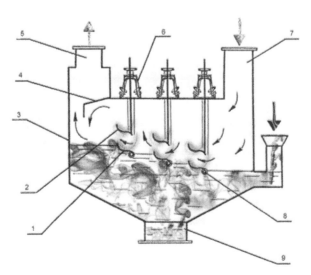

FIGURE 10.1   General view of a rotoklon: lower 1 and the overhead 2 lobes; a body 3; the labyrinth drip pan 4; connecting pipes for an entry 7 and an exit 5 gases; screw jacks 6; an axis 8; a connecting pipe for a drain of slurry 9.

The mentioned structural features do not allow using correctly available solutions on hydrodynamics of dust-laden gas flows for a designed construction. In this connection, for the well-founded exposition of the processes occurring in the apparatus, there was a necessity of realization of experimental research studies.

Experiments were conducted on the laboratory-scale plant "rotoklon" and presented in Figure 10.2.

The examined rotoklon had three slotted channels speed of gas with gas speed up to 15 m/s. At this speed, the rotoklon had a hydraulic resistance 800 Pa. Working in such regime, it supplied efficiency of trapping of a dust with input density 0.5 g/nm$_3$ and density 1,200 kg/m$_3$ at level of 96.3 per cent (Usmanova, 2013).

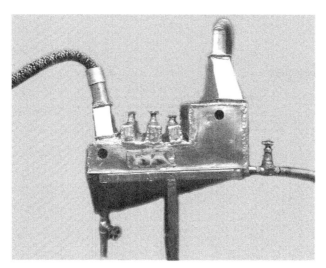

FIGURE 10.2   Experimental installation "rotoklon": In the capacity of modeling system, air and a dust of talc with a size of corpuscles $d = 2 - 30$ μm, white black and chalk have been used. The apparatus body was filled with water on level $h_g = 0.175$ m.

Cleanliness level of an airborne dust mix was determined by a direct method (Kouzov, 1987; and Shcwidki, 2002). On direct sections of the pipeline before and after the apparatus, the mechanical sampling of an airborne dust mix was made. After determination of matching operational mode of the apparatus, gas test was performed by means of tubes for mechanical. On tubes for research studies various diameter tips have been installed.

Full trapping of the dust contained in taken test of an airborne dust mix was made by an external filtering draws through mixes with the help of calibrates electro-aspirator EA-55 through special analytical filters AFA-10 which were put in into filtrating cartridges. The selection time was fixed on a stop watch, and speed—the rotameter of electro-aspirator EA-55.

Experimental installation for a mechanical sampling at definition of a dustiness of gas on a method of an external filtering is shown in Figure 10.3.

Dusty gas is selected from the flue by an in taking tube 3 and filtrated through filter AFA-10 fixed in a cartridge 4. The cleared gas from a cartridge arrives in the glass diaphragm 9 connected to the differential manometer 15 and further in the blowing machine 13. Directly ahead of a glass diaphragm are measured gas temperature (the thermometer 8) and its rarefaction (a mercury manometer 16).

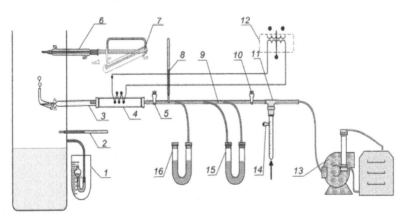

FIGURE 10.3   Experimental research of speed and stream dustiness.

On a line from a glass diaphragm to the blowing machine, there is a tee-joint 11 which is connected to an atmosphere a rubber tube supplied with a crimped lock 14. By means of this crimped lock control speed of selection of gas from the flue, false air in the blowing machine is changing. On a section from a cartridge to a glass diaphragm install a crimped lock 5, using which, it is possible to change a rarefaction at a glass diaphragm. In-taking tube 3 and a cartridge 4 are connected to the transformer 12, with an output voltage of 12 volt. At the expense of it they have an electric heating.

Speed of gas is measured in the flue by a tube 6 which is connected to the micromanometer 7. The temperature and pressure (rarefaction) of gas are measured in the flue accordingly by the thermometer 2 and a manometer 1.

Dust gas mix gained by dust injection in the flue by means of the metering screw conveyer batcher presented in Figure 10.4. Application of the batcher with varying productivity has given the chance to gain the set dust load on an entry in the apparatus.

FIGURE 10.4   The metering screw conveyor batcher of a dust.

The water discharge is determined by its losses on transpiration and with deleted slurry. The water drain is made in the small portions from the loading pocket supplied with a pressure lock. Gate closing implements sweeping recompression of air in the gate chamber, opening—a depressurization. Small-level recession is sweepingly compensated by a top up through a connecting pipe of feeding into of a fluid. At periodic drain of the condensed slurry, the water discharge is determined by consistency of slurry and averages to 10 g on 1 m_3 air, and at fixed drain the charge does not exceed 100–200 g on 1 m_3 air. Filling of a rotoklon with water was controlled by means of the level detector. Maintenance of a fixed level of water has essential value as its oscillations involve appreciable change as efficiency and productivity of the device.

10.3   DISCUSSION OF RESULTS OF EXPERIMENT

In a rotoklon process of interaction of gas, liquid, and solid phases in which result the solid phase (dust), finely divided in gas, passes in a fluid are realized. In the process of hydrodynamic interaction of phases in the apparatus, it is possible to disjoint sequentially proceeding stages on the following: fluid acquisition by a gas flow on an entry in the contact device; fluid subdivision by a fast-track gas flow in the contact channel; concretion of dispersion particles by liquid drops; branch of drops of a fluid from gas in the labyrinth drip pan.

The inspection of the observation port shows that all channels are filled by foam and water splashes. Actually this effect is caused by a retardation of a flow at an end wall, is characteristic only for a stratum which directly is bordering on to glass. Slow-motion shot consideration allows to install a true flow pattern. It is visible that the air jet as though itself chooses the path, being aimed to be punched in the shortest way through water. Blades standing sequentially under existing conditions restrict air jet extending, forcing it to make sharper turn that, undoubtedly, favors to separation. Functionality of all dust trap depends on the efficiency of acquisition of a fluid a gas flow—without fluid acquisition will not be supplied effective interacting of phases in contact channels and, hence, qualitative clearing of gas of a dust will not be attained. Thus, fluid acquisition by a gas flow at consecutive transiting of blades of an impeller is one of the defined stages of hydrodynamic process in a rotoklon.

Fluid acquisition by a gas flow can be explained by the presence of interphase turbulence which is advanced on an interface of gas and liquid phases. Conditions for origination of interphase turbulence are presence of a gradient of speeds of phases on boundaries, difference of viscosity of flows, and interphase surface tension.

10.4   THE ESTIMATION OF EFFICIENCY OF GAS CLEANING

The quantitative assessment of efficiency of acquisition in apparatuses of shock-inertial type with inner circulation of a fluid is expedient for conducting by means of a parameter $n = L_z/L_g$, m$_3$/m$_3$ equal to the ratio of volumes of liquid and gas phases in contact channels and characterizing the specific charge of a fluid on gas irrigating in channels. Obviously that magnitude n will be determined, first of all, by speed of a gas flow on an entry in the contact channel. The following important parameter is fluid level on an entry in the contact channel which can change the cross-section of the channel and influence speed of gas:

$$\frac{\vartheta_g}{S_g} = \frac{\vartheta_g}{bh_k - bh_l} - \frac{\vartheta_g}{b(h_k - h_l)} \tag{10.1}$$

where S_g is the cross section of the contact channel, m$_3$; b is the channel width, m; h_K is the channel altitude, m; h_1 is the fluid level, m.

Thus, for the exposition of acquisition of a fluid a gas flow in contact channels of a rotoklon, it is enough to gain the following relation experimentally.

$$n = f(\vartheta_g \cdot h_l) \tag{10.2}$$

As it has been installed experimentally, efficiency of trapping of corpuscles of a dust in many respects depends on the size of drops of a fluid: with decrease of a size of drops, the dust clearing efficiency raises. Thus, the given stage of hydrodynamic interaction of phases is rather important. For calculation of average diameter of the drops organized at transiting of blades of an impeller, the empirical relation is gained:

$$d = \frac{467 \cdot 10^3 \sqrt{\sigma}}{\vartheta_o} + 17{,}869 \cdot \left(\frac{\mu_l}{\sqrt{\rho_l \sigma}} \right)^{0{,}68} \frac{L_l}{L_r} \tag{10.3}$$

where υ is the relative speed of gases in the channel, mps; σ is the factor of a surface tension of a fluid, N/m; ρ_1 is the fluid density, kg/m$_3$; μ_1 is the viscosity of a fluid, the Pascal/s; L_1 is the volume flow of a fluid, m$_3$/s; L_g is the volume flow of gas, m$_3$/s.

The offered formula allows to also consider together with physical performances of phases and agency of operating conditions.

In Figure 10.5, design values of average diameter of the drops organized at transiting of blades of an impeller, from speed of gas in contact channels and a gas-specific irrigation are introduced. In calculation, values of physical properties of water were accepted at temperature 20°C: $\rho_1 = 998$ kg/m$_3$; $\mu_1 = 1.002 \cdot 10_{-3}$ N $\cdot$S/m$_2$, and $\zeta = 72.86 \, 10_{-3}$ N/m

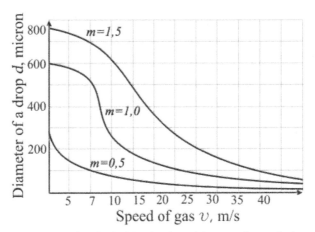

FIGURE 10.5   Computational relation of a size of drops to flow velocity and a specific irrigation.

The gained relations testify that the major operating conditions on which the average size of drops in contact channels of a rotoklon depend, speed of gas flow υ and the specific charge of a fluid on gas irrigating n are. These parameters determine hydrodynamic structure of an organized water gas flow.

Separation efficiency of gas bursts in apparatuses of shock-inertial act can be discovered only on the basis of empirical data on particular constructions of apparatuses. Methods of calculations which are put into practice are based on an assumption about possibility of linear approximation of dependence of separation efficiency from diameter of corpuscles in a likelihood-logarithmic coordinate system.

Calculations on a likelihood method are executed under the same circuit design, as for apparatuses of dry clearing of gases (Shcwidki, 2002).

Shock-inertial sedimentation of corpuscles of a dust occurs at flow of drops of a fluid by a dusty flow; therefore, the corpuscles possessing inertia continue to move across the curved stream-lines of gases, the surface of drops attain and are precipitated on them.

Efficiency of shock-inertial sedimentation $\eta_{\text{и}}$ is function of following dimensionless criterion:

$$\eta_{\text{è}} = f\left(\frac{m_p}{\xi_c} \cdot \frac{\vartheta_p}{d_0}\right) \qquad (10.4)$$

where m_p is the mass of a precipitated corpuscle; υ_p is the speed of a corpuscle; ξ is the factor of resistance of driving of a corpuscle; d_0 is the diameter a midelev of cross section of a drop.

For the spherical corpuscles which driving obey the law the Stokes, this criterion looks like the following:

$$\frac{m_p \vartheta_p}{\xi_c d_0} = \frac{1}{18} \cdot \frac{d_r^2 \vartheta_p \rho_p C_c}{\mu_g d_0} \tag{10.5}$$

Complex $d_p^2 \vartheta_p \rho_p C_c / \left(18 \mu_g d_0\right)$ is a parameter (number) of the Stokes

$$\eta_{\grave{e}} = f\left(Stk\right) = f\left(\frac{d_p^2 \vartheta_p \rho_p C_c}{18 \mu_g d_0}\right) \tag{10.6}$$

Thus, efficiency of trapping of corpuscles of a dust in a rotoklon on the inertia model depends primarily on performance of a trapped dust (a size and density of trapped corpuscles) and operating conditions major of which is the speed of a gas flow at transiting through blades of impellers.

In Figure 10.6, results of calculation of a dust clearing efficiency by means of formula (10.6) are shown. For various sizes of dust, the increase in general efficiency of dust separation with increase in number of Stokes is observed.

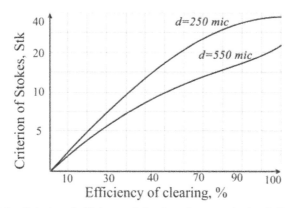

FIGURE 10.6   Relation of efficiency of clearing of gas to criterion StK

On the basis of the observed inertia of model, the method of calculation of a dust clearing efficiency in scrubbers with inner circulation of a fluid is developed.

The basis for calculation on this model is formula (10.6). To understand the calculation, it is necessary to know disperse composition of a dust, density of corpuscles of a dust, viscosity of gas, speed of gas in the contact channel, and the specific charge of a fluid on gas irrigating.

Calculation is conducted in the following sequence:

- By formula (10.3) determine an average size of drops D_0 in the contact channel at various operating conditions.
- By formula (10.4) count the inertia parameter of the Stokes for each fraction of a dust.
- By formula (10.5) calculate the fractional values of efficiency η for each fraction of a dust.
- General efficiency of dust separation is determined by formula (10.6), %.

The observed inertial model in detail characterizes the physics of the processes flowing in contact channels of a rotoklon.

10.5   COMPARISON OF EXPERIMENTAL AND COMPUTATIONAL RESULTS

On the basis of analyzing the gained results of research studies of general efficiency of a dust separation, it is necessary to underscore that in a starting phase of activity of a dust trap for all used in research studies dust separation high performances, components from 93.2 per cent for carbon black to 99.8 per cent for a talc dust are gained. Difference of general efficiency of trapping of various types of a dust originates because of their various particle size distributions on an entry in the apparatus, and also because of the various forms of corpuscles, their dynamic wettability, and density. The gained high values of general efficiency of a dust separation testify to correct the selection of constructional and operation parameters of the studied apparatus and indicate its suitability for use in engineering of a wet dust separation.

As shown in Figures 10.7 and 10.8, the relation of general efficiency of dust separation on speed of a mixed gas and fluid level in the apparatus will well be agreed to design data that confirms an acceptability of the accepted assumptions.

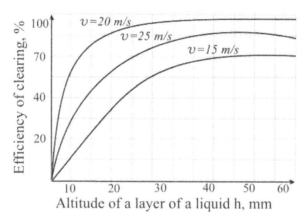

FIGURE 10.7   Relation of efficiency of clearing of gas to irrigating liquid level.

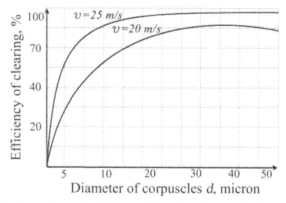

FIGURE 10.8   Dependence of efficiency of clearing of gas on the size of corpuscles and speed of gas.

The results of research studies on trapping of various dusts in a rotoklon with adjustable sinusoidal are shown in Figure 10.9. The given research studies testify a high performance of trapping of corpuscles for low dust with their various moistening abilities. From these drawings by fractional efficiency of trapping it is obviously visible, what even for corpuscles a size less than 1 μm (which are most difficultly trapped in any type of dust traps) installations efficiency considerably above 90 per cent. Even for the unwettable sewed type of white black, general efficiency of trapping was more than 96 per cent. Naturally, as for the given dust trap lowering of fractional efficiency of trapping at decrease of sizes of corpuscles less than 5 μm, however not such sharp, as or other types of dust traps is characteristic.

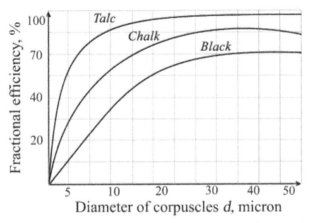

FIGURE 10.9   Fractional efficiency of clearing of corpuscles of various dusts.

As appears from Figure 10.10, schedule presented in drawing, dependence of general efficiency of a dust separation on the inertia parameter of Stokes will well be coordinated with design data that confirms an acceptability of our assumptions.

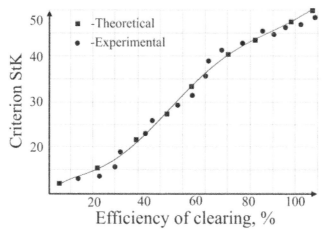

FIGURE 10.10   General efficiency of clearing of gas emissions depending on the criterion of stokes.

10.6   CONCLUSIONS

1. The new construction of the rotoklon is developed in order to solve the problem of effective separation of a dust from a gas flow. In the introduced apparatus, water admission to contact zones implements as a result of its circulation in the device.

2. Experimentally, it is shown that fluid acquisition by a gas flow at consecutive transiting of blades of an impeller is one of the defined stages of hydrodynamic process in a rotoklon.

3. The theoretical concepts are confirmed by immediate measured data of efficiency value of shock-inertial sedimentation of dispersion particles in a rotoklon. The gained desing relationships allow to size up the contribution of characteristics of a collected dust (a size and density of collected corpuscles). Also, it is possible to size up the contribution of operating conditions important such of which is speed of a gas stream at passage through blades of impellers.

4. Good convergence of results of scalings on the gained relationships with the data which are available in the technical literature and this experiment confirms an acceptability of the accepted assumptions.

The formulated leading-outs are actual for intensive operation wet-type collectors in which the basic gear of selection of corpuscles is the gear of the inertia dust separation.

KEYWORDS

- Contact channels
- Dust
- Irrigation
- Rotoklon
- Separation
- The efficiency of gas purification

REFERENCES

1. Kouzov, P. A.; Malgin, A. D.; and Skryabin, G. M.; Clearing of Gases and Air of a Dust in the Chemical Industry. St.-Petersburg: Chemistry; **1993.**
2. Kouzov, P. A.; Bases of the Analysis of Disperse Composition Industrial a Dust. Leningrad: Chemistry; **1987.**
3. Pirumov, A. I.; Air Dust Removal. Moscow: Engineering Industry; **1974.**
4. Shcwidki, V. S.; and Ladygichev, M. G.; Clearing of Gases. The Directory. Moscow: Heat Power Engineering; **2002.**
5. Straus, V.; Industrial Clearing of Gases. Moscow: Chemistry; **1981.**
6. Shctokman, E. A.; Air Purification. Moscow: Publishing House ASV; **1998.**
7. Usmanova, R. R.; Zhernakov, V. S.; and Panov, A. K.; "Rotoklon a Controlled Sinusoidal Blades" R. F. Patent 2317845, February 27, **2008.**
8. Usmanova, R. R.; Zaikov, G. E.; Stoyanov, O. V.; and Klodziuska, E.; Research of the Mechanism of Shock-Inertial Deposition of Dispersed Particles from Gas Flow the Bulletin of the Kazan Technological University. **2013,** *16(9),* 203–207.
9. Uzhov, V. H.; Valdberg, A. J.; and Myagkov, B. I.; Clearing of Industrial Gases of a Dust. Moscow: Chemistry; **1981.**
10. Valdberg, A. J.; and Lebedjuk, G. K.; Wet-Type Collectors of Shock and Centrifugal. Moscow: Petrochemistry; **1981.**
11. Vatin, N. I.; and Strelets, K. I.; Air Purification by Means of Apparatuses of Type the Cyclone Separator. St. Petersburg; **2003.**

CHAPTER 11

MATHEMATICAL MODELING OF TRAFFIC OF DISPERSION PARTICLES IN BLADE IMPELLERS

CONTENTS

11.1 INTRODUCTION

From the literary data it follows that known builds of scrubbers with internal circulation of a liquid work in a narrow range of change of speed of gas. Because of low speeds of gas in contact channels known apparatuses have the big gabarits. These defects, and also a weak level of scrutiny of processes occurring in apparatuses, absence of reliable methods of their calculation complicate working of new rational builds of wet-type collectors of the given type and their wide implementation in manufacture.

The quantitative estimation of efficiency of process of sedimentation on a liquid surface can be executed on the basis of consideration of physical model of interacting of counter streams. With some assumptions, it is possible to present a physical model of interacting of counter streams as specific targets about an inleakage of a stream of a gas stream on a liquid surface. Using a method of conformal mappings, it is possible to gain the dependence merging co-ordinates of points on a surface of sedimentation with speeds [1–3].

The following was central tasks, which were put by working out of a new build of a rotoklon with internal circulation of a liquid:

• to create the apparatus with a wide range of change of operating conditions and a wide scope, including for clearing of gases of the basic industrial assemblies of a finely divided dust;• to gain the differential equation of motion of corpuscles with which, help was possibly to define paths of their motion in the field of an inleakage of a gas stream on a liquid surface, and also to count limiting sizes of corpuscles, which can be precipitated on a liquid surface [4–8].

11.2 STATEMENT OF PROBLEM OF HYDRODYNAMICS OF DISCONTINUOUS PHASE

Research of this or that problem of gas kinetics of an aerosol or motion of a single corpuscle is initiated from problem formalization, that is, from its record in a mathematical aspect—in the form of system of the equations.

In some difficult cases, this stage of research is an independent problem (Belousov, 1988).

Methodically the stage of a statement of problem develops of following stages:

1. Sampling of physical model of process. At this stage is accepted any idealization, the schematization, etc. For example, instead of an imperfect gas is accepted any of classical models: model of perfect fluid (gas) or viscous fluid model. Within the limits of the set model, it is possible to inject the assumption of external affecting in addition.

2. Sampling of a frame of reference in which motion and a condition of the studied medium is presented. For example, Eulerian method when the coordinate are motionless, and moves medium concerning it, or system Lagranzha when the coordinate are connected with moving a corpuscle, and motion is studied concerning this corpuscle.

3. Record of the universal equations of thermodynamics, mechanics, electrodynamics, etc. The equations can be in the differential or integrated form.

4. The additional equations characterizing conditions of uniqueness:
   • indicate the area occupied with medium, and a time interval;
   • conditions at infinity;
   • singular points in medium (charges, point sources);
   • entry conditions;
   • boundary conditions.

5. Introduction of the simplifications connected with decrease of number of independent variables, for example:
   • motion installed is expelled a time t;
   • motion plane-parallel is expelled one co-ordinate, the coordinate system is chosen so that speeds of corpuscles were parallel a plane x–y;
   • motion potential, and a liquid incompressible. This simplification gives the chance to reduce a problem to search only one unknown.

6. Introduction of the simplifications connected with linearization of the equations.

As a rule, the basic equations of motion, energy, etc., nonlinear as required functions are included into the basic equations and in the equations of boundary conditions generally nonlinearly.

Linearization of the equations not only simplifies the equations and their solution, but also results thereto that it is possible to use superposition principle of solutions. This principle consists that the sum of several private solutions is the equation common decision whereas for system of the nonlinear equations, the sum of private solutions is not the common decision.

To gain the analytical solution of the equations of gas kinetics of aerosols rather difficultly in connection with essential complication as differential equations of motion and energy, and equations boundary and entry conditions. It leads to necessity to induct simplification, to replace exact, but difficult communications between magnitudes approached, but more simple. At the heart of such simplifications lie transition from usual variables to generalized and determination of principles or modeling conditions.

Method of scale transformations use when all basic equations of processes are known. This method allows to attain an identity of dimensionless forms of the equations. The generalized variables or a similarity parameter thus come to light.

Method of the analysis of dimensions of a quantity apply when observed processes have no mathematical description, and the relationships characterizing processes in the most general terms are known only.

11.3   DEVELOPMENT OF MOTION OF CORPUSCLES IN THE ROTOKLON

The great interest represents consideration of processes proceeding in apparatuses of impact-sluggish act. The overall picture of interacting of a gas stream with a liquid surface in impact-sluggish apparatuses is resulted on Figure 11.1 and can be presented on the basis of works (Hertzian, 1975; and Gurevich, 1979) and visual examination.

In the course of an inleakage of a gas stream on a liquid surface the dint (plunge basin) in a liquid, which depth depends on a dynamic pressure of a gas stream is formed. Over the range, speed of a gas stream from 5 to $10°$m/s depth of a dint is insignificant. Wave formation on a liquid surface has quiet character and fades on some distance from an axis of an accumulating gas stream. At speed of a gas stream more than $10°$m/s wave formation is accompanied by partial break-down of drops of a liquid by a backward jet of a gas stream. At the further increase in speed of a gas stream intensity of a tip leakage increases.

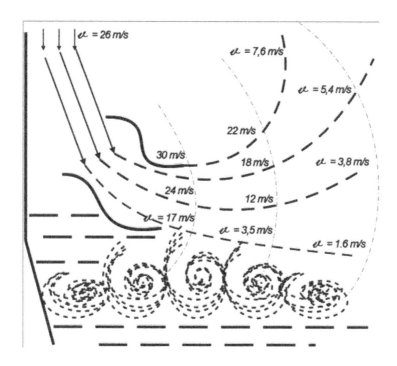

FIGURE 11.1   Stream-lines in impeller shovels.

Pre-dominantly, the tip leakage occurs from a dint lateral surface. The size oscillated from a surface of a liquid of drops makes $300 \div 600$ mm (Kutateladze, 1976).

Depth of the dint formed at an inleakage of a stream on a surface of a liquid, is defined from expression (Kusenkov, 1971):

$$\frac{h}{d} = n\sqrt{A_r}$$

where h–distance from blades of an impeller to liquid level in the apparatus, mm;

d–diameter of the upstream end, mm;

n–a dimensionless coefficient;

A_r–the Archimedes number defined by formula:

$$A_r = \frac{\rho V_g^{\,2}}{gd\rho_l} \tag{11.1}$$

where V_g–speed of gas on an entry in an impeller, mps.

Diameter of a dint is defined by formula [1–4]:

$$\frac{D_v}{d} = 1 + 0,67(\frac{h}{d})^{0,85}$$

(11.2)

Speed of a backward jet

$$Vo = Vg \cdot \frac{d^2}{D_v^2 - d^2}$$

(11.3)

If the stream of final width "a" accumulates on a stream of infinite width the current is presented by the equation:

$$\frac{dW}{d\xi} = \frac{8\alpha\xi}{\pi(\xi+1)(\xi-1)^3}$$

(11.4)

$\xi = dW/dZ$–complex speed of a current;

$W = f + i\Psi$–complex potential;

$Z = x + iy$–complex coordinate of a current.

After separation of variables and integration, the Eq. (11.4) becomes:

$$W = \frac{\alpha}{\pi}\left[\ln\frac{\xi+1}{\xi-1} - 2\frac{\xi}{(\xi-1)^2}\right]$$

(11.5)

By means of the Eq. (11.4) and substitutions

$$\frac{dZ}{d\xi} = \frac{dZ}{dW} \cdot \frac{dW}{d\xi} = \frac{1}{\xi} \cdot \frac{dW}{d\xi}$$

(11.6)

the interconnection between ζ and Z is defined

$$\frac{dZ}{d\xi} = \frac{8a}{\pi(\xi+1)(\xi-1)^2}$$

(11.7)

or

$$Z = \frac{a}{\pi}\left[\ln\frac{\xi-1}{\xi+1} + 2\frac{\xi-2}{(\xi-1)^2} + 4\right]$$

(11.8)

The integration constant in the Eq. (11.8) is chosen in such a manner that in a critical point 0 complex speed of a current $\zeta = 0$. The Eqs. (11.5) and (11.8) define current function ψ, potential of speed f and present a field of speeds in a zone of interaction of a stream of a gas stream with a liquid surface.

After transformations and reduction to a dimensionless aspect, it is possible to gain the dependences connecting,

$$Z = \frac{\pi \cdot Z}{a} = x + iy = \frac{\pi \cdot x}{a} + i\frac{\pi \mp \cdot y}{a}$$

$$W = \frac{\pi W}{a} \text{ and } \zeta:$$

$$Z = -\left[W + \frac{4}{(\xi - 1)^2} - 4 \right] \tag{11.9}$$

where Z–dimensionless complex co-ordinate;
W–dimensionless complex potential.

Having expression for X and Y, it is possible, using the differential equation of motion of corpuscles to define paths of their motion in the field of an inleakage of a gas stream on a liquid surface. Generally, the equation of motion of a corpuscle in a gas stream looks like:

$$m_r \frac{d\vec{V}_r}{\tau} = \vec{F} \tag{11.10}$$

where m_r–weight of a corpuscle;
V_r–a vector of absolute speed;
F–resultant vector of forces acting on a corpuscle.

For area of usability of the formula of Stokes that is for corpuscles with $d \leq 70$ mcm, the Eq. (11.10) can be written down in an aspect:

$$\frac{\pi d_r^3}{6} p_r \frac{d\vec{V}_r}{d\tau} = \frac{\pi d_r^3}{6} p_r \vec{g} - 3\pi\mu d_r (\vec{V} - \vec{V}_r) \tag{11.11}$$

The Eq. (11.11) is convenient to lead to a dimensionless aspect having injected following designations:

$$V_{r0} = \frac{V_r}{V_\infty}; V_0 = V_0 = \frac{\vec{V}}{V_\infty}; \tau_0 = \tau \frac{V_\infty}{a}$$

then

$$\frac{d\vec{V}_{r0}}{d\tau_0} = \frac{\vec{g}}{g} F_r - \frac{1}{St}(\vec{V}_0 - \vec{V}_{r0}) \qquad (11.12)$$

where $F_r = \dfrac{ga}{V_\infty^2}$ – a Froude number.

$$St = \frac{d_r^2 p_r V_\infty}{18\mu a} \text{–Criterion of Stokes.}$$

Using correlation:

$$V_{r0x} = \frac{dx}{d\tau_0}; V_{r0y} = \frac{dY}{d\tau_0} = \frac{d^2x}{d\tau_0^2}; \frac{dV_{r0y}}{d\tau_0} = \frac{d^2Y}{d\tau_0^2}$$

It is possible to gain system of the differential equations in coordinates X and Y corpuscles defining path:

$$\frac{d^2x}{d\tau_0^2} = -F_r - \frac{1}{St}(V_{0x} - \frac{dx}{d\tau_0})$$

$$\frac{d^2y}{d\tau_0^2} = -\frac{1}{St}(V_{0y} - \frac{dy}{d\tau_0}) \qquad (11.13)$$

Calculations on the Eqs. (11.11–11.13) allow to define sizes of corpuscles, which can be precipitated on a liquid surface, and also build critical paths of corpuscles (Figure 11.2).

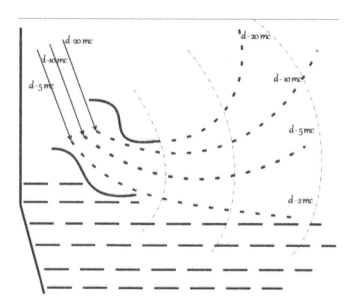

FIGURE 11.2   Paths of corpuscles of a dust.

Analogous method calculation of paths of corpuscles for of Stokes field forces of resistance of a gaseous fluid to motion of corpuscles can be executed.

The resulted dependences refer to cases when the positive allowance between target cross-section of blades of an impeller and a liquid surface is almost equal to null. Under such circumstances along with sedimentation of firm corpuscles there is intensive enough generating of a liquid from its surface, which then is separated from a gas stream by means of special devices.

11.4   CONCLUSIONS

1. The physical model of interacting of counter streams, which can be observed as specific targets about an inleakage of a stream of a gas stream on a liquid surface, is presented. The dependences connecting co-ordinates of points on a surface of sedimentation with speeds of a stream are gained. Calculations on the devised model

allow defining sizes of corpuscles, which can be precipitated on a liquid surface, and also build critical paths of corpuscles.

2. Good convergence of results of scalings on the gained relationships with the data, which is available in the technical literature and own experiments confirms an acceptability of approximating assumptions.

KEYWORDS

- **Archimedes number**
- **Criterion of stokes**
- **Streamlines**
- **The backward jet**
- **The Froude number**
- **The path of corpuscles**
- **The rotoklon**

REFERENCES

1. Belousov, V. V.; Theoretical Bases of Process of Clearing of Gas Emissions. M: Metallurgy;24, 92–111, **1988.**
2. Valdberg, A. J.; and Lebedjuk, G. K.; Wet-Type Collectors Impact-Sluggish, Centrifugal and Injector Acts. M.: Petrochemistry; **1981.**
3. Hertzian, M. И.; Kirsanov, H. C.; and Gordon, M.; Effect of separate factors on efficiency of trapping of a dust in percussion scrubbers. *Works GIN Tsvet Meta.* 1975, *36,* 112–117, *(in Russian).*
4. Gurevich, M. A.; The Theory of Streams of Perfect Fluid. M.: Science; **1979.**
5. Kusenkov, B. A.; Research of wet-type collectors with internal circulation of water. *Water Supply Sanitary Tech.* 1971, *11,* 19–23, *(in Russian).*
6. Kutateladze, S. S.; and Styrikovich, M. A.; Hydrodynamics of Gas-Liquid Systems. M: Energy; **1976.**
7. Milne-Thomson, L. M.; Theoretical Hydrodynamics. M.: World; **1964.**
8. Richkov, C. P.; and Savelyev, J. A.; Application of a rotoklon "Ural" in the industry. Safety of Work in the Industry. **1982,** *8,* 43–45, *(in Russian).*

CHAPTER 12

AERODYNAMIC PROFILING OF BLADES OF AN IMPELLER

CONTENTS

12.1   INTRODUCTION

The basic defect of apparatuses with internal circulation of a liquid is dependence of extent of a dust separation and expenses of energy of a dusty stream through a liquid. Considerable power inputs on gas clearing lead to search of new original design and technological solutions. A new problem arose in creation of such dedusters at which the organization of traffic of a stream in the apparatus would promote decrease in maintenance costs at clearing of dusty streams without detriment to efficiency of clearing. Actual a hydrodynamics direction is creation of such hydrodynamic circumstances in dedusters, which allows to raise considerably efficiency of processes and thus to lower a water resistance. It is the momentous direction of experimental and modeling researches for the purpose of understanding of the physical mechanism of a whirlwind intensification and optimization of processes in separation and dust-collecting plants. Application of computing technologies and software packages, which allows to count with accuracy comprehensible to practice hydrodynamic characteristics in turbulent eddy flows for a development testing and designing of industrial devices, including dust removal, allowing to avoid necessity of expensive full-scale tests [1–12].

12.2   MODELLING OF TRAFFIC AND GAS SEPARATION

It is expedient to apply aerodynamically rational roll forming of blades of an impeller to water resistance decrease.

Giving to shovels of an impeller sinusoidal a profile allows to eliminate breakaway a stream breakaway on edges. Thus, there is a flow of an entrance section of a profile of blades with the big constant speed and increase in ricochets from a shaped part of blades in terms of which it is possible to predict insignificant increase in efficiency of clearing of gas [13–15]. The analytical model of shaped blades is represented on Figure 12.1.

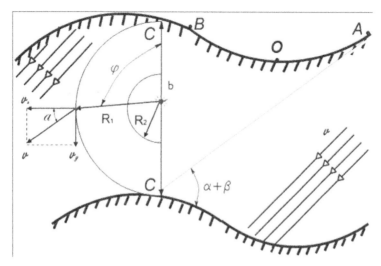

FIGURE 12.1 The analytical model of shaped blades of an impeller.

The stream of dusty gas through impeller shovels (Figure 12.1) can be schematized figure a flat incompressible flow, which approaches to impellers at an angle $\alpha+\beta$ within a speed υ and at turnabout edges of blades loses dust corpuscles, rebounding from blades and entrained scrub solution [16–20]. The cleared gas comes off от edges of blades, forming in channels between them detached flow with constant pressure. Boundary lines detached flow can be modeled as a line of rupture of tangential speeds in an ideal (nonviscous) liquid and to use for current calculation known methods of the theory of streams (Gurevich, 1979).

According to this theory on boundary lines detached flow pressure p_c and speed υ_c are constant and connected by Bernoulli number with pressure p_x and speed υ_x in a relative wind:

$$p_c/p_r + 0,5\upsilon_c^2 = p_1/p_2 + 0,5\upsilon_1^2 = p*/p_g \qquad (12.1)$$

$p*$ – pressure of the retarded stream, roughly equal (to within hydraulic losses in the upstream end) to pressure p_0 gas to an entry in impellers.

In perfect fluid model if to neglect a friction and mixing, the gas stream passing through shovels, represents a row of parallel streams in width b, leaking with a speed υ_2 at an angle $\alpha>0$. Perfect fluid parameters in cross-sections $C–A$ and $C–C$ are connected by a continuity equation:

$$v_1 S \sin \alpha_1 = v_2 b, \tag{12.2}$$

in Bernoulli number

$$p_1/p_r + v_1^2/2 = p_2/p_\Gamma + v_1^2/2 = p*/p_\Gamma \tag{12.3}$$

and momentum equation (the equation of pulses) in a projection to a heading of blades

$$-v_1 \rho_\Gamma Q_K \cos(\alpha_1 + \beta) + \rho_1 F_K \sin \beta = v_2 \rho_\Gamma Q_K \cos(\alpha_2 - \beta) + \rho_2 F_K \sin \beta, \tag{12.4}$$

Where a Q_k-volume gas rate through an impeller; F_k–the area of a face-to-face surface of one shovel.

In terms of that $Q_K = v_1 F_K \sin \alpha_1$, from system of the Eqs. (12.3) and (12.4) we will gain:

$$v_2/v_1 = (1/\sin \beta)\left[\sin \alpha_1 \cos \cdot (\alpha_2 - \beta) + \sqrt{\sin^2 \alpha_1 \cos^2 (\alpha_2 - \beta) + \sin \beta \cdot \sin(2\alpha_1 + \beta)}\right] \tag{12.5}$$

On the schedule (Figure 12.2), results of calculations under this formula for various α_2 are resulted. Value of an angle a_2, in turn, depends from a and β, and also on distance between shovels b

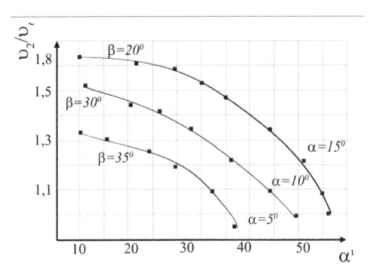

FIGURE 12.2   Angle effect α' for the speed v_2/v_1 at Various α and β.

Width of a stream after its turn between impeller shovels define on a relationship:

$$\frac{b}{S} = \frac{v_1}{v_2}\sin\alpha_1 = \frac{\sin\alpha_1\sin\beta}{\sin\alpha_1 + \sin(\alpha_1 + \beta)}. \qquad (12.6)$$

Definition of fields of speeds of gas in shovels of an impeller by means of the mentioned theory of streams (Gurevich, 1979) represents labor-intensive process. For the simplified solution of a problem of calculation of separation of a dust, it is possible to be restricted to the approached model of a gas stream considering the basic regularity, installed at more exact calculation of fields of speeds (Stepans, 1986). Such computational model is represented on Figure 12.1. The gas stream between shovels at stream turn is modeled by a ring with internal radius R_1 and external radius $R_2 > R_1$. A stream angle of rotation define a geometrical relationship.

$$\Psi_{max} = \pi - (\alpha + \beta)$$

Assuming that gas is in regular intervals distributed on a face-to-face surface of a shovel, and disregarding a thickness of a shovel, speed of a relative wind of gas can be defined from the charge equation:

$$v_1 = G_j /(\rho_g F_p \sin\alpha),$$

Where G_M–a mass flow rate of the cleared gas; F_p–the square of a face-to-face surface of blades.

After turn, the stream has certain width в and speed v_2, which find from a continuity condition

$$v_2 = v_1 (S\sin\alpha)/b.$$

At stream turn cross-section on internal boundary line speed is equal in everyone v_2 and changes as a function of the radial co-ordinate r under the law of a potential circular stream:

$$v = v_2 R_1 /r.$$

Velocity distribution comparison in cross-sections O–A and O–B (Figure 12.1) according to the simplified analytical model shows that the simplified velocity distribution is close to more exact, differing in the beginning in smaller bearing, and then in the big. After cross-section *O-B*difference becomes less.

Width в *a* stream after turn define according to dependence Eq. (12.6) with introduction of a correction index of expansion of a stream *a*>1:

$$B = aS \frac{\sin \beta}{1 + \sin(\alpha + \beta)/\sin \alpha}.$$

The factor *a* compensates a difference between the design and experimental values, caused by viscosity of gas and turbulent expansion of a stream. A numerical value of factor *a* choose on the basis of comparison of results of calculation and matching experiments by definition factors of the dropping of a dust in a rotoklon.

Sampling of factor *a* essentially influences design values of partial and general factors of the dropping of a dust. This results from the fact that speeds of gas at stream turn are proportional to speed v_2, i.e., inversely proportional to value в. Thus, securing separation of firm corpuscles the centrifugal force is proportional to a square of tangential speed making them, which depends on speed of gas. The results of calculations compared with experimental data show that at $a=1.50 \div 1.55$ design values of the general factors of the dropping of a dust converge well with the experiment.

At calculations, according to the analytical model 12.1, in an entrance part gas-liquid a stream free boundary lines of a stream are replaced with shaped channels of blades of an impeller. In a blow point O,*a* relative wind, it is atomized and flows round curvilinear surface *AB*, thus speed attains the maximum value ϑ_{max} and remains a constant on length of all section. In cross-section*C*, *a* stream, it is possible to observe as homogeneous, moving with a speed ϑ, and impeller shovels have a final thickness, the breakaway of a stream from impeller blades does not occur. Airflow, the stream flow is carried out without pressure loss to critical cross-section *C–C* after which static pressure in a stream initiates to drop.

Radiuses R_2 and R_1 stream boundary lines at stream turn define from a continuity condition

$$bv_2 = \int\limits_{R_1}^{R_2} v dr = \int\limits_{R_1}^{R_2} \left(v_2 R_1 / r \right) dr .$$

As a result of integration

$$R_2 = R_1 \exp(b/R_1).$$ (12.7)

From geometrical relationships (seeFigure 12.1)

$$R_2 = S \sin \beta - R_1 \cos(\alpha + \beta).$$ (12.8)

From the Eqs.(12.8) and (12.7)

$$R_1 = \frac{S \sin \beta}{\exp(b/R_1) + \cos(\alpha + \beta)}.$$ (12.9)

This equation solve a method of iterations. In the capacity of root R_0 of initial approximation, it is possible to use value R_1, which derive provided that $(R_2 - R_1)$=в

Then from the Eq.(12.9)

$$R_0 = \frac{S \sin \beta - b}{1 + \cos(\alpha + \beta)}.$$

It is expedient to apply a method of Aitkena-Steffensena to an acceleration of convergence of iterative process under the following circuit design. If an approximation by iteration order (i=0; 1; 2; 3; 4...) any multiple of 3, compute

$$R_i = \varphi(R_{i-1})$$

where φ (R) the first side of an Eq. (12.9).

Thus, test of convergence of iterations the usual

$$\vec{\Delta R_i} = |(R_i - R_{i-1})/R_i| \le \vec{\Delta}_{R(min)}$$ (12.10)

If i any multiple of 3, apply the formula

$$R_i = R_{i-3} - \frac{(R_{i-2} - R_{i-3})^2}{R_{i-1} - 2R_{i-2} + R_{i-3}}.$$ (12.11)

In this case, iterations complete solution when one of two test of convergence is secured. The first sign, usual, by formula (12.10). Another sign – value of a denominator close enough to null in the Eq. (12.11)

$$f(R_1)_i = (R_{i-1} - 2R_{i-2} + R_{i-3})/S \leq f(R_1)_{min}.$$

In check, computation satisfactory convergence of iterations is gained at $f(R)_{min} = 0.015$.

$$f(R_1)_i = (R_{i-1} - 2R_{i-2} + R_{i-3})/S \leq f(R_1)_{min}.$$

12.3 EXPERIMENTAL RESEARCH OF AEROHYDRODYNAMIC CIRCUMSTANCES

Carry out an experiment on the laboratory facility "rotoklon" (See Figure 12.3).

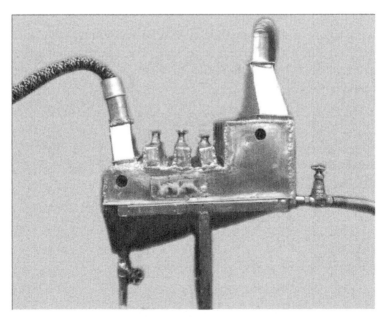

FIGURE 12.3 Experimental installation "rotoklon."

In a rotoklon, three pairs of the blades having profile of a sinusoid are installed. Blades can be controlled for installation of their position. Depending on cleanliness level of an airborne dust flow the lower lobes by means of handwheels are installed on an angle defined by operational mode of the device. The rotoklon is characterized by presence of three channels, a formation of the overhead and bottom blades and in each following on a run of gas the channel the bottom blade is installed above the previous. Such arrangement promotes a gradual entry of a water gas flow in slotted channels and thereby reduces the device hydraulic resistance. The arrangement of an input part of lobes on an axis with a capability of their turn allows creating a diffusion reacting region. Sequentially, slotted channels mounted in a diffusion zone equipped with a rotation angle lobe, a hydrodynamic zone of intensive wetting of corpuscles of a dust. In process of flow moving through the fluid-flow curtain, the capability of multiple stay of corpuscles of a dust in hydrodynamically reacting region is supplied that considerably raises a dust clearing efficiency and ensures functioning of the device in broad bands of cleanliness level of a gas flow. The observed build of the apparatus is protected by the patent of the Russian Federation.

The examined rotoklon had three slotted channels speed of gas with gas speed up to 15mps. At this speed, the rotoklon had a hydraulic resistance 800 Pa. Working in such regime, it supplied efficiency of trapping of a dust with input density $0.5g/nm^3$ and density $1,200kg/m^3$ at level of 96.3 percent (Usmanova, 2013).

In the capacity of modeling system, air and a dust of talc with a size of corpuscles $d=2\div30$ a micron, white black and a chalk have been used. The apparatus body was filled with water on level $h_g=0.175m$.

Hydrodynamically, it is expedient to profile impeller shovels in terms of various values of speeds within slot-hole channels. Thus, the gas-dispersed stream can be swing on any necessary angle with the least hydraulic losses (Bussrojd, 1975; and Kutateladze, 1976).

As have shown results of experiment, angle change α from 30° to 45° promotes increase in a flow area of stream R_2-R_1 at veering of traffic of gas and to increase b after its turn. The stream flow area b increases on the average in 1.55 times. Thus, speed of gas at passage through impeller shovels decreases that carry decrease tangential components v_y speeds of dispersion particles. The centrifugal force of inertia responsible for branch of corpuscles from a gas stream and defining an increment by radial components v_x of speed of corpuscles, thus also decreases. Expansion of stream, R_2-R_1 promotes increase in a mechanical trajectory of corpuscles at separation, and it in turn conducts to considerable growth of secondary ablation of a dust.

The stream of dusty gas passing through shovels of an impeller, it is possible to schematize a flat incompressible flow, which approaches to a lattice at an angle α within a speed v and at turnabout edges of blades loses dust corpuscles, rebounding from blades. The cleared gas comes off edges of blades, forming in slot-hole channels between them detached flows with approximately constant pressure. Boundary lines detached flow can be observed as a line of rupture of tangential speeds in an ideal (nonviscous) liquid. Definition of fields of speeds of gas at passage through impeller shovels inconveniently enough. The calculation problem can be simplified, if at research of process of separation of a dust to observe the approached model of the gas stream considering the cores installed at more exact calculation behavior of fields of speeds of gas-dispersed medium.

Thus, if design data of blades of an impeller are set, it is possible to count a gas velocity distribution at a flow its gas-dispersed stream. Knowing speeds of gas in various points of a gas stream, it is possible to define forces of an aerodynamic resistance on which mechanical trajectories of firm corpuscles in slot-hole channels depend.

If for a usual rotoklon efficiency of trapping of a dust was small and sharp depended from fluid level, filled in the apparatus in the presence of impeller blades efficiency trapping in all velocity band of air was high and did not depend on liquid level in the apparatus (See Figures 12.4–12.6).

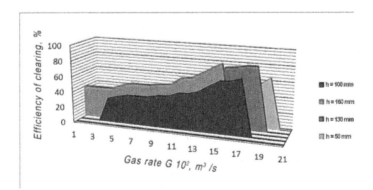

FIGURE 12.4 Dependence of efficiency of clearing of gas on air consumptions and irrigating liquid without blades.

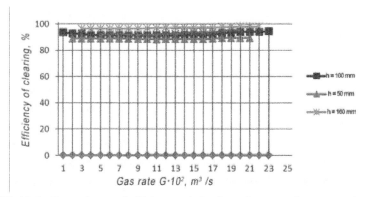

FIGURE 12.5 Dependence of efficiency of clearing of gas on air consumptions and irrigating liquid with sinusoidal shovels

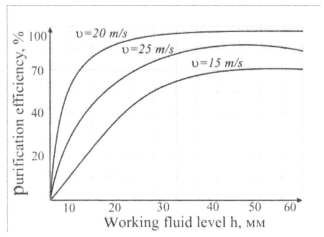

FIGURE 12.6 Dependence of efficiency of clearing of gas on liquid level at various speed of gas.

12.4 RESEARCH OF HYDRODYNAMIC CHARACTERISTICS IN PROGRAM ANSYS-14 CFX

Judging by descriptions in the literature, the bending of blades pursues the aim to excite a centrifugal force and to use them for separation of corpuscles within the bent channel an impeller. The great value concerning a dust separation is attributed also to curtains of drops and the splashes, twice crossing an air stream path.

Considering a low water resistance of sinusoidal blades of an impeller with an intensive twisting of streams of air in their each element, it is possible to draw a conclusion that similar shovels have good prospects for the solution warmly and mass-transfer problems in system gas-liquid (Aksenov, 1996; Launder, 1974; and Vasquez, 2000).

The physical picture of interacting of air and liquid in impeller shovels is presented on Figure 12.7.

FIGURE 12.7   Interacting of gas and irrigating liquid.

Computer modeling has been applied to studying of hydrodynamic characteristics in the apparatus in program ANSYS-14 CFX. For calculation, the method of final elements was used. The current of gas without a discrete phase has been simulated. Power and thermal affecting of corpuscles on a current of a gas phase negligibly small. This assumption is true in that case when the mass fraction of corpuscles in a two-phase stream does not exceed 30 percent (Crowe, 1998). The Numerical analysis of a current of gas with great dispatch-vortical apparatus is reduced to the solution of system of mean field Navier–Stokes equations on Reynolds. For short circuit of gas-dynamic Navier—Stokes equations, it was used standard k-e turbulence model. Besides turbulence and heat exchange model, it was used volume of fluid(VOF) for modeling with a free surface of water and a stream of air (Hirt, 1981).

Boundary conditions have been chosen the following: in the capacity of gas has been used air, liquids on which air-water goes, temperature on an entry–300K, pressure of gas on an exit hardly above atmospheric 101425 Pa, intensity of turbulence on an entry and an apparatus exit–5 percent.

The first investigation phase was definition of a role of blades of an impeller. For this purpose, comparative researches by efficiency of a dust separation in one apparatus have been conducted at two builds:

Usual rotoklon (Figure 12.8); a rotoklon with the shovels of an impeller offered in the present work (Figure 12.9).

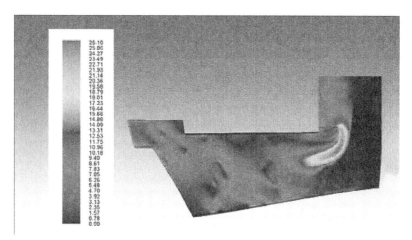

FIGURE 12.8   Field of speeds without impeller blades.

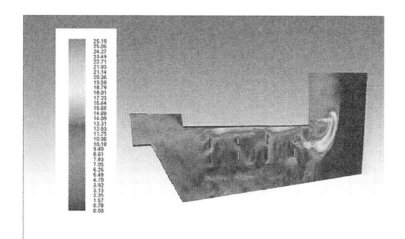

FIGURE 12.9   Field of speeds with impeller shovels.

From the presented drawings, it is visible that the gas stream in the apparatus with impeller blades in an injection zone has numerous uniform eddyings. Formation of a turbulent trace Behind impeller shovels is formed a turbulent trace as a result of a breakaway of a boundary layer from its surface, and also growth of frequency of turbulent pulsations.

Besides, intensity of turbulence that the gas-cleaning installation is in the momentous parameter gains in strength. Field of speeds (see Figure 12.9) shows that with installation of sinusoidal blades of an impeller the uniform distribution of speeds in the apparatus is observed more.

Hence, it is possible to draw a conclusion that impeller shovels reinforce turbulence of cleared gas in a zone of injection of the apparatus that theoretically should affect increase in efficiency of separation of finely divided impurity. By means of modeling, it has been found out that with since speed of bringing gas of 15mps, the intensive carryover of liquid is initiated and there is a necessity to make good the loss. It is installed that at raise of level of a liquid, considerably increases not only a rotoklon water resistance, but also there are turbulent pulsations of pressure with high magnitudes of differences.

12.5   CONCLUSIONS

1.   It is experimentally installed that the aerodynamic roll forming of blades of an impeller on a sinusoid allows lowering a device water resistance considerably. Thus, there is an increase in efficiency of clearing of gas tanks to a flow of an entrance section of a profile of blades with the big constant speed and to increase in ricochets from a shaped part of blades.
2.   Selection constructional and the operating conditions, securing high efficiency of a dust separation at small factor of a water consumption, allows to recommend such dedusters for implementation in the industry.

KEYWORDS

- **A blade profile**
- **Hydrodynamics**
- **Modeling**
- **The irrigating liquid**
- **The rotoklon**
- **The stream**
- **Turbulent pulsations**

REFERENCES

1. Aljamovskij, A. A.; SolidWorks 2007/2008. Computer Modeling in Engineering Practice. SPb.: BhV-Peterburg;**2008.**
2. Bussrojd, R.; Current of Gas with the Weighed Corpuscles. M.: The World;**1975.**
3. Gurevich, M. I.; The Theory of Streams of Perfect Fluid. M: The Science;**1979.**
4. Egorov, N. N.; Gas Cooling in Scrubbers. M.: Chemistry Publishing House;**1954.**
5. Kutateladze, S. S.; and Styrikovich, M. A.; Hydrodynamics of Gas-Liquid Systems. M: Energy;**1976.**
6. The Patent 2317845 Russian Federations, The Rotoklon with Adjustable Sinusoidal Guide Vanes. Usmanova, R. R.; Zhernakov, V. S.; and Panov, A. K.; February 27.**2008.**
7. Stepans, G. J.; and Zitser, I. M.; The Inertia Air Cleaners. M.: "Engineering Industry";**1986.**
8. Margopin, E. V.; and Prihodko, V. P.; Perfection of Production Engineering of Wet Clearing of Gas Emissions at AluminiumFactories. M.: Colour. Metal TheInformation;**1977.**
9. Crowe, C.; Sommerfield, M.; and Yutaka Tsuji. Multiphase Flows with Droplets and Particles. CRC Press;**1998.**
10. Aksenov, A. A.; Dyadkin, A. A.; and Gudzovsky, A. V.; Numerical Simulation of Car Tire Aquaplaning Computational Fluid Dynamics 96.Eds. Desideri, J. A.; Hirsch, C.; Tallec, P. Le.; Pandolfi, M.; Periaux, J.; John Wiley&Sons;**1996,** 815–820 p.
11. Dukowicz, J. K. A.; Particle-fluid numerical model for liquid sprays. *J. Comput. Phys.* **1980,** *35,* 229–253.
12. Harlow, F. H.; and Welch, J. E.; Numerical calculation of time-dependent viscous incompressible.*Flows Fluid Free Surface Phys. Fluids.* **1965,** *8,* 2182–2187.
13. Hinze, J. O.; Turbulence. New York: McGraw-Hill;**1975.**
14. Hirt, C. W.; and Nicholls, B. D.; Volume of Fluid (VOF) method for dynamical free boundaries.*J. Comput. Phys.* **1981,** *39,* 201–225.
15. Launder, B. E.; and Spalding, D. B.; The numerical computation of turbulent flows. *Comp. Meth. Appl. Mech. Eng.***1974,***3,* 269–289.

16. Menter, F. R.; Multiscale model for turbulent flows in 24th fluid dynamic conference. *Am. Inst. Aero. Astro.* **1993**.

17. Menter, F. R.; Two-equation eddy-viscosity turbulence models for engineering applications. *AIAA J.* **1994**,*32(8)*.

18. Menter, F. R.; and Esch, T.; Advanced Turbulence Modelling in CFX CFX Update Spring **2001**,*20*, 4–5.

19. Usmanova, R. R.; Zaikov, G. E.; Stoyanov, O. V.; and Klodzinska, E.; Research of the Mechanism of Shock-Inertial Deposition of Dispersed Particles from Gas Flow.The Bulletin of the Kazan Technological University;**2013**,*16(9)*, 203–207 p. *(in Russian)*.

20. Vasquez, S. A.; and Ivanov, V. A.; A phase coupled method for solving multiphase problems on unstructured meshes. In: Proceedings of ASME FEDSM'00: ASME 2000 Fluids Engineering Division Summer Meeting. Boston; June **2000**.

CHAPTER 13

EXPERIMENTAL RESEARCH AND CALCULATION OF BOUNDARY CONCENTRATION OF AN IRRIGATING LIQUID

CONTENTS

13.1   INTRODUCTION

One of trends of development of wet-type collectors is creation of apparatuses of intensive operation with high carrying capacity on a gas phase that is connected with favorable decrease in gabarits of installations. In these conditions, owing to high relative speed of traffic of liquid and gas phases, work upon effect of a dust separation mechanisms: the inertia and direct capture of corpuscles. Such process is realized in impact-sluggish dedusters to which it is possible to refer to the investigated apparatus[1–5].

Authors Jarzkbski and Giowiak(1977), analyzing work of an impact-sluggish deduster have installed that in the course of a dust separation defining role is played by the phenomenon of the inertia collision of a dust with water drops. Efficiency of allocation of corpuscles of a dust decreases together with growth of sizes of the drops oscillated in the settling space. In case of a generating of drops compressed air, their magnitude is defined by equation Nukijama and Tanasawa(1939). From the equation follows that drops that more than above value of viscosity of a liquid phase.

Viscosity growth can call reduction of efficiency of a dust separation. In the literature practically, there are no data on effect of viscosity of a trapping liquid on dust separation process. Therefore, one of the purposes of our work was revealing of effect of viscosity of a liquid on efficiency of a dust separation[6–15].

13.2   CURRENT STATE OF A PROBLEM

It is possible to admit that for certain three-phase systems gas–a liquid–a solid boundary concentration of a dust can appear too big that will directly affect too big extent of circulation, which is difficult for realizing in the conditions of commercial operation of wet-type collectors.

The problem formulation leans on following rules. In the conditions of full circulation of a liquid, at constant geometrical sizes of a deduster, it is possible to secure with a constancy of operational parameters is a relative speed of traffic of a liquid and an aerosol, concentration of a dust in gas, a superficial tension of a liquid, or an angle of wetting of a dust. Concentration of a dust growing in a time in a liquid conducts to unique essential change–to increase in its viscosity. After excess of certain concentration suspension loses properties of a Newtonian fluid. Deduster working conditions at full circulation of a liquid are approached to what can be gained in the periodical regime when at maintenance fresh water is not inducted into dust-collecting plant. Collected in the apparatus, the dust detained by a liquid, compensates volume losses of the liquid necessary on moistening of passing gas and its ablation. In the literature, there are no works, theoretically both experimentally presenting effect of viscosity and effect of rheological properties of slurry on efficiency of a dust separation. As it seems to us, a motive is that fact that in the capacity of operating fluid water is usually used, and dedusters work, predominantly, at constant temperatures. Simultaneously, at use of partial circulation certain level of concentration of a dust in a liquid is secured. In turn, accessible dependences in the literature specify in insignificant growth of viscosity of slurry even at raise of its concentration for some percent.

The reasoning's proving possibility of effect of viscosity of slurry on efficiency of a dust separation, it is possible to refer to as on the analysis of the basic mechanisms influencing sedimentation of corpuscles on an interphase surface, and on conditions of formation of this developed surface of a liquid. Transition of corpuscles of a dust from gas in a liquid occurs, mainly, as a result of the inertia affecting, effect of "sticking" and diffusion. Depending on type of the wet-type collector of a corpuscle of a dust deposit on a surface of a liquid, which can be realized in the form of drops, moving in a stream of an aerosol, the films of a liquid generated in the apparatus, a surface of the gas vials formed in the conditions of a barbotage and moistened surfaces of walls of the apparatus.

In the monography(Egorov, 1954) effect of various mechanisms on efficiency of sedimentation of corpuscles of a dust on a liquid surface is widely presented. The description of mechanisms and their effect on efficiency of a dust separation can be found practically in all monographies, for example (Kutateladze, 1976; and Rist, 1987), concerning a problem wet clearings of gas emissions of gases. In the literature of less attention, it is given questions of formation of surfaces of liquids and their effect on efficiency of a dust separation.

Observing the mechanism of the inertia act irrespective of a surface of the liquid entraining a dust, predominantly, it is considered that for hydrophilic types of a dust collision of a part of a solid with a liquid surface to its equivalently immediate sorbtion by a liquid, and then immediate clearing and restoration of the Surface of a liquid for following collisions. In case of a dust badly moistened, the time necessary for sorbtion of a corpuscle by a liquid, can be longer, than a time after which the corpuscle will approach to its surface. Obviously, it is at the bottom of decrease in possibility of a retardation of a dust by a liquid because of a recoil of the corpuscle going to a surface, from a corpuscle which are on it. It is possible to consider this effect real as in the conditions of a wet dust separation with a surface of each fluid element impinges more dust, than it would be enough for monolayer formation. Speed of sorbtion of corpuscles of a dust can be a limiting stage of a dust separation.

Speed of sorbtion of a corpuscle influences not only its energy necessary for overcoming of a surface tension force, but also and its traverse speed in the liquid medium, depending on its viscosity and rheological properties. Efficiency of dust separation Kabsch(2003) connects with speed of ablation of a dust a liquid, having presented it as weight ms, penetrating in unit of time through unit of a surface and in depth of a liquid as a result of a collision of grains of a dust with this surface:

$$r = \frac{m_s}{A \cdot t}$$

Giving to shovels of an impeller sinusoidal a profile allows to eliminate breakaway, a stream breakaway on edges. Thus, there is a flow of an entrance section of a profile of blades with the big constant speed and increase in ricochets from a shaped part of blades in terms of which it is possible to predict insignificant increase in efficiency of clearing of gas.

Speed of linkage of a dust a liquid depends on physicochemical properties of a dust and its ability to wetting, physical, and chemical properties of gas and operating fluid, and also concentration of an aerosol. Wishing to confirm a pushed hypothesis, Kabsch(2003) conducted the researches concerning effect of concentration speed of linkage of a dust by a liquid. The increase in concentration of a dust in gas called some increase in speed of linkage, however to a lesser degree, than it follows from linear dependence.

The cores for techniques of a wet dust separation of model Semrau, Barth'a, and Calvert'a do not consider effect of viscosity of slurry on effect of a dust separation. In-process Pemberton'a(Slinna, 1981), it is installed that in case of sedimentation of the corpuscles, which are not moistened on drops, their sorbtion in a liquid is obligatory, and their motion in a liquid submits to principle Stokes'a.

The traverse speed can characterize coefficient of resistance to corpuscle motion in a liquid, so, and a dynamic coefficient of viscosity of a liquid. Possibility of effect of viscosity of a liquid on efficiency of capture of corpuscles of a dust a drop by simultaneous Act of three mechanisms: the inertia, "capture" the semiempirical equation Slinna(1981) considering the relation of viscosity of a liquid to viscosity of gas also presents.

In general, it is considered that there is a certain size of a drop (Stepans, 1986) at which optimum conditions of sedimentation of corpuscles of a certain size are attained, and efficiency of subsidence of corpuscles of a dust on a drop sweepingly decreases with decrease of a size of these corpuscles.

Jarzkbski[9]), analyzing work of an impact-sluggish deduster have installed that in the course of a dust separation defining role is played by the phenomenon of the inertia collision of a dust with water drops. Efficiency of allocation of corpuscles of a dust decreases together with growth of sizes of the drops oscillated in the settling space, in case of a generating of drops compressed air, their magnitude is defined by equation Nukijama and Tanasawa(1939) from which follows that drops to those more than above value of viscosity of a liquid phase. Therefore, viscosity growth can call reduction of efficiency of a dust separation.

The altitude of a layer of the dynamic foam formed in dust-collecting plant at a certain relative difference of speeds of gas and liquid phases, decreases in process of growth of viscosity of a liquid (Shwidkiy, 2002) that calls decrease in efficiency of a dust separation, it is necessary to consider that the similar effect refers to also to a layer of an intensive barbotage and the drop layer partially strained in dust removal systems.

Summarizing, it is possible to assert that in the literature practically there are no data on effect of viscosity of a trapping liquid on dust separation process. Therefore, one of the purposes of our work was extraction of effect of viscosity of a liquid on efficiency of a dust separation.

13.2.1 THE PURPOSE, ASSUMPTIONS, AND AREA OF RESEARCH

The conducted researches had a main objective acknowledging of a hypothesis on existence of such boundary concentration of slurry at which excess the overall performance of the dust removal apparatus decreases.

The concept is devised and the installation, which is giving the chance to implementation of planned researches is mounted. Installation had systems of measurement of the general and fractional efficiency and typical systems for measurement of volume flow rates of passing gas and water resistances. The device of an exact proportioning of a dust, and also the air classifier separating coarse fractions of a dust on an entry in installation is mounted. Gauging of measuring systems has secured with respective repeatability of the gained results.

13.3 LABORATORY FACILITY AND TECHNIQUE OF CONDUCTING OF EXPERIMENT

Laboratory facility basic element is the deduster of impact-sluggish act–a rotoklon c adjustable guide vanes (Usmanova, 2008) (see Figure 13.1). An aerosol gained by dust introduction in the pipeline by means of the batcher. Application of the batcher with changing productivity has given the chance to gain the set concentration of a dust on an entry in the apparatus.

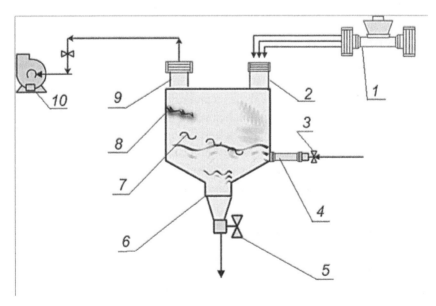

FIGURE 13.1   The laboratory facility.

Have been investigated a dust, discriminated with the wet ability (a talcum powder the ground, median diameter is equal δ50=25mm, white black aboutδ50=15mmsolubilityin water of 10⁻³ percent on weight (25°C) and a chalk powder).

The gas-dispersed stream passed shovels of an impeller 7 in a working zone of the apparatus, whence through the drip pan 8 cleared, was inferred outside. Gas was carried by means of the vacuum pump 10, and its charge measured by means of a diaphragm 1. A Gas rate, passing through installation, controlled, changing quantity of air sucked in the pipeline before installation. Gas differential heads were measured by a manometer. Concentration of a dust in entering and getting out gases measured by means of the analogous systems presented in Chapter 10. The composition of each system includes group of three sondes mounted on vertical sections of pipelines, on distance about 10 diameters from the proximal element changing the charge. The taken test of gas went on the measuring fine gauge strainer on which all dust containing in test separated. For this purpose used fine gauge strainer. In the accepted solution have applied system of three measuring sondes, which have been had in pipelines so that in the minimum extent to change a regime of passage of gas and to select quantity of a dust necessary for the analysis. The angle between directions of deducing of sondes made 120°, and their ends placed on such radius that surfaces of rings from which through a sonde gas was sucked in, were in one plane. It has allowed to scale down a time of selection of test and gave average concentration of a dust in gas pipeline cross-section.

Fractional composition of a dust on an entry and an exit from the apparatus measured by means of analogous measuring systems, Chapter 10.

For definition of structurally mechanical properties of slurry viscosity RV-8 (see Figure 13.2) has been used. The viscosity gauge consists of the internal twirled cylinder (rotor) (r=1.6cm) and the external motionless cylinder (stator) (r=1.9cm), having among themselves a positive allowance of the ring form with a size 0.3 see the Rotor is resulted in twirl by means of the system consisting of the shaft, a pulley (To=2.23cm), filaments, blocks, and a cargo. To the twirl termination apply a brake. The twirled cylinder has on a division surface on which control depth of its plunging in slurry.

FIGURE 13.2   Measurement of viscosity of slurry.

The gained slurry in number of 30sm3 (in this case the rotor diving depth in sludge makes 7sm) fill in is carefully the washed out and dry external glass, which put in into a slot of a cover and strengthen its turn from left to right. After that again remove the loaded cylinder that on a scale of the internal cylinder precisely to define depth of its' plunging in sludge. Again fix a glass and on both cups put the minimum equal cargo (on 1), fix the spigot of a pulley by means of a brake, and reel up a filament, twirling a pulley clockwise. It is necessary to watch that convolutions laid down whenever possible in parallel each other.

Install an arrow near to any division into the limb and, having hauled down a brake, result the internal cylinder in twirl, fixing a time during which the cylinder will make 4–6 turns. After the termination of measurements fix a brake and reel up a filament. Measurement at each loading spend not less than three times. Experiences repeat at gradual increase in a cargo on 2 gr. until it is possible to fix a time of an integer of turns precisely enough. After the termination of measurements remove a glass, drain from it sludge, wash out water, from a rotor sludge drain a wet rag then both cylinders are dry wiped and leave the device in the collected aspect.

After averaging of the gained data and calculation of angular speed, the schedule of dependence of speed of twirl from the enclosed loading is under construction, Viscosity is defined by formula (Shwidkiy, 2002)[15]:

$$\eta = \frac{(R_2^2 - R_1^2)Gt}{8\pi^2 L R_1^2 R_2^2 L} \tag{13.1}$$

or
$$\eta = \frac{kGt}{L} \tag{13.2}$$

13.4 DISCUSSION OF RESULTS OF RESEARCHES

For each dust used in researches dependence of general efficiency of a dust separation on concentration of slurry and the generalizing schedule of dependence of fractional efficiency on a corpuscle size is presented. Other schedule grows out of addition fractional efficiency of a dust separation for various, presented on the schedule, concentration of slurry. In each case, the first measurement of fractional efficiency is executed in the beginning of the first measuring series, at almost pure water in a deduster.

Analyzing the gained results of researches of general efficiency of a dust separation, it is necessary to underline that in a starting phase of work of a rotoklon at insignificant concentration of slurry for all used in dust researches components from 93.2 percent for black to 99.8 percent for a talc powder are gained high efficiency of a dust separation. Difference of general efficiency of trapping of various types of a dust originates because of their various fractional composition on an entry in the apparatus, and also because of the various form of corpuscles, their dynamic wet ability and density. The gained high values of general efficiency of a dust separation testify to correct selection of constructional and operational parameters of the studied apparatus, and specify in its suitability for use in techniques of a wet dust separation.

The momentous summary of the spent researches was definition of boundary concentration of slurry various a dust after which excess general efficiency of a dust separation decreases. Value of magnitude of boundary concentration, as it is known, is necessary for definition of the maximum extent of recirculation of an irrigating liquid. As presented in Figures 13.3–13.6 schedules, dependence of general efficiency of a dust separation on concentration of slurry, accordingly, for a powder of talc, a chalk, and white black is available possibility of definition of such concentration.

Boundary concentration for a talcum powder–36 percent, white black–7 percent, a chalk–of 18 percent answer, predominantly, to concentration at which slurries lose properties of a Newtonian fluid. The conducted researches give the grounds to draw deductions that in installations of impact-sluggish type where the inertia mechanism is the core at allocation of corpuscles of a dust from gas, general efficiency of a dust separation essentially drops when concentration of slurry answers such concentration at which it loses properties of a Newtonian fluid. As appears from presented in Figures 13.3–13.6 dependences, together with growth of concentration of slurry above a boundary value, general efficiency of a dust separation decreases, and the basic contribution to this phenomenon small corpuscles with a size less bring in than 5mm. To comment on the dependences presented in Figures, than 5°mm operated with criterion of decrease in efficiency of a dust separation of corpuscles sizes less at the further increase in concentration of slurry at 10 percent above the boundary. Taking it in attention, it is possible to notice that in case of allocation of a dust of talc growth of concentration of slurry from 36 to 45 percent calls reduction of general efficiency of a dust separation from 98 to 90 percent at simultaneous decrease in fractional efficiency (Figure 13.3) of allocation of corpuscles, smaller 5mm from η=93 percent to η=65 percent (Figure 13.4).

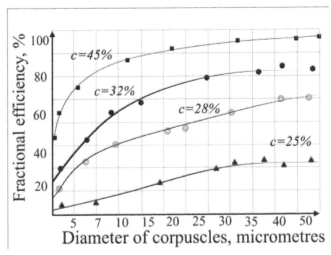

FIGURE 13.3 Dependence of fractional efficiency on diameter of corpuscles of a talcum powder and their concentration in a liquid.

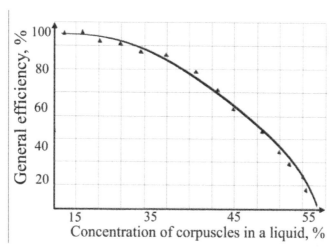

FIGURE 13.4   Dependence of general efficiency of concentration in a liquid of corpuscles of a talcum powder.

Analogously for white black: growth of concentration from 7 to 20 percent calls falling of fractional efficiency from $\eta=65$ percent to $\eta=20$ percent, for a chalk: growth of concentration from 18 to 30 percent calls its decrease from $\eta=80$ percent to $\eta=50$ percent. Most considerably, decrease in fractional efficiency of a dust separation can be noted for difficultly moistened dust–white black (about 50%)(Figures 13.5 and 13.6).

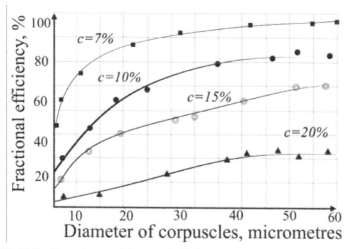

FIGURE 13.5   Dependence of fractional efficiency on diameter of corpuscles of white black and their concentration in a liquid.

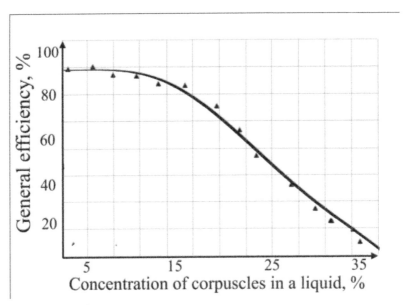

FIGURE 13.6 Dependence of general efficiency on concentration in a liquid of corpuscles of white black.

Thus, on the basis of the analysis set forth above, it is possible to assert that decrease in general efficiency of a dust separation at excess of boundary concentration of slurry is connected about all by decreasing ability of system to detain small corpuscles. Especially, it touches badly moistened corpuscles. It coincides with a hypothesis about updating of an interphase surface. Updating of an interphase surface can be connected also with difficulties of motion of the settled corpuscles of a dust deep into liquids, i.e., with viscosity of medium.

The analysis of general efficiency of a dust separation, and, especially, a talcum powder and white black powder specifies that till the moment of achievement of boundary concentration efficiency is kept on a fixed level. In these boundary lines, simultaneously with growth of concentration of slurry dynamic viscosity of a liquid, only this growth grows is insignificant–for talc, for example, to 2.7×10^{-3}Pa sec. At such small increase in viscosity of slurry, the estimation of its effect on efficiency of a dust separation is impossible. Thus, it is possible to confirm, what not growth of viscosity of slurry (from 1 to 2.7′10-3Pa·sec), and change of its rheological properties influences decrease in efficiency of a dust separation.

The method of definition of boundary extent of circulation of a liquid in impact-sluggish apparatuses is based on laboratory definition of concentration of slurry above, which it loses properties of a Newtonian fluid. This concentration will answer concentration of operating fluid, which cannot be exceeded if it is required to secure with a constant of efficiency of a dust separation.

In the conditions of spent researches, i.e., constant concentration of an aerosol on an entry in the apparatus, and at the assumption that water losses in the apparatus because of moistening of passing air and, accordingly, ablation in the form of drops, is compensated by volume of the trapped dust, the water discharge parameter is defined directly from the recommended time of duration of a cycle and differs for various types of a dust. Counted its maximum magnitude is in the interval 0.02–0.05l/m3, i.e., close to the magnitudes quoted in the literature.

For periodical regime dedusters, this concentration defines directly a cycle of their work. In case of dedusters of continuous act with liquid circulation, the maximum extent of recirculation securing maintenance of a fixed level of efficiency of a dust separation, it is possible to count as the relation:

$$r = \frac{Q_{cir}}{Q_{ir}} \qquad (13.3)$$

where Q_{cir}–the charge of a recycling liquid, m3/h; Q_{ir}–the charge of an inducted liquid on an irrigation, m3/h.

Assuming that all dust is almost completely trapped on a liquid surface, it is possible to write down a balance equation of weight of a dust as:

$$G \cdot (c_{on} - c_{in}) = L \cdot c_{on} \qquad (13.4)$$

where limiting concentration of a dust, g/m3

Then in terms for calculation of extent of recirculation, it is possible to present Eq. (13.4) formula as:

$$r = \frac{G \cdot c_{on}}{L \cdot c_r} \qquad (13.5)$$

13.5 CONCLUSIONS

1. Excess of boundary concentration of slurry at which it loses properties of a Newtonian fluid, calls decrease in efficiency of a dust separation.
2. At known boundary concentration cr, it is possible to define boundary extent of recirculation of the irrigating liquid, securing stable efficiency of a dust separation.
3. Magnitude of boundary concentration depends on physical and chemical properties of system, a liquid-solid and changes over a wide range, from null to several tens percent. This magnitude can be defined now only laboratory methods.
4. Decrease in an overall performance of the apparatus at excess of boundary concentration is connected, first of all, with reduction of fractional efficiency of trapping of small corpuscles with sizes less than 5mm.
5. On the basis of observations of work of an investigated deduster, it is possible to assert that change of viscosity of an irrigated liquid influences conditions of a generating of an interphase surface and, especially, on intensity of formation of a drop layer.
6. Selection constructional and the operating conditions, securing high efficiency of a dust separation at small factor of water consumption, allows to recommend such dedusters for implementation in the industry.

KEYWORDS

- **Boundary concentration**
- **Fractional efficiency**
- **Interfacial area**
- **Recirculation**
- **The irrigating liquid**
- **The rotoklon**
- **Viscosity**

REFERENCES

1. Valdberg, A. J.; Wet-Type Collectors Impact-Sluggish, Centrifugal and InjectorActs. M.: Petrochemistry;1981.
2. GOST 21235-75 Talc and a talcum powder the ground. Specifications.
3. Egorov, N. N.; Gas Cooling in Scrubbers. M.: Chemistry Publishing House;1954.
4. Kutateladze, S. S.; and Styrikovich, M. A.; Hydrodynamics of Gas-Liquid Systems. M.: Energy;1976.
5. Kabsch-Korbutowicz, M.; andMajewska-Nowak, K.; Removal of atrazine from water by coagulation and adsorption. Environ. Protect. Eng. 2003,29(3), 15–24.
6. Kitano, T.; and Slinna, T.; "An empirical equation of the relative viscosity of polymer melts filled with various inorganic fillers."Rheologica.Acta.1981, 20(2), 207.
7. Margopin, E. V.; and Prihodko, V. P.; Perfection of Production Engineering of Wet Clearing of Gas Emissions at AluminiumFactories. M: Colour Metal the Information;1977.
8. Nikolaev, N.A.;Rectificative columns with vortical direct-flow steps.TheoryBasis Chem. Technol.1970,2, 261–264,(inRussian).
9. Jarzkbski, L.; and Giowiak.; An. OchrvmyIbrodItr.1977.
10. Nukijama, S.; and Tanasawa, Y.; Trans. Soc. Mech. Eng. Japan.1939,5.
11. The Patent 2317845 Russian Federations.In: The Rotoklon with Adjustable Sinusoidal Guide Vanes.Usmanova,R. R.;Zhernakov, V. S.; and Panov, A. K.; February 27, 2008.
12. Ramm, V. M.; Absorption of Gases. M: Chemistry;1976.
13. Rist, R.; Aerosols: Introduction in the Theory.M.:The World;1987.
14. Stepans, G. J.; and Zitser, I. M.; The Inertia Air Cleaners. M: "Engineering Industry";1986.
15. Shwidkiy, V. S.; Purification of Gases. In: Handbook.Shwidkiy, V. S.;Moscow:ThermalPower;2002.

CHAPTER 14

TECHNICAL AND ECOLOGICAL ASSESSMENT OF SAMPLING OF SYSTEM OF CLEARING OF GAS

CONTENTS

14.1   INTRODUCTION

Colossal scales of industrial human activity have led to the big positive transformations to the world—to creation of powerful industrial and agricultural potential, wide development of all types of transport, an irrigation, and melioration big ground area the squares, to creation of systems of an artificial climate. At the same time has sharply worsened air state. The further decline air state can lead to far-reaching negative aftereffects for mankind. Therefore wildlife management, its protection against pollution became one of the major global problems (Smiths, 1979; Feodors, 1972; and Tsygankov, 1979).

One of consequences of technogenic effect on a circumambient in a number of the countries now is the appreciable decline of a condition of a free air.

Most largest catalytic unit (million т. In a year) aerosphere global pollutions form CO_2 ($2 \cdot 10^4$), WITH (200), SO_2 (150), NO_x (50), hydrogen sulphide [1–3]

Clearing of a gas stream fathom branch from it or transformations into the harmless form of the pollutants, which are thrown out in an aerosphere together with a gas stream. By pollution air masses can be transferred on the big distances and essentially influence a condition of an aerosphere and health of the person [3–5].

In particular, a year accumulation in aerosphere CO_2, the global increase of temperature (greenhouse effect) can call 0.4 percent occurring to intensity. Conversion in aerosphere SO_2, NO_x, and other the analogous nature of emissions can come to the end with formation of aid fogs and to fall of the acid rains (snow) calling corrosion of many inorganic materials, and as to oppression and destruction various background objects—florae and faunae [5–7].

This condition causes the rigid demands shown to industrial atmospheric emissions and dirt content in a free air. Performance of these demands is monitored by special services of the factories, and also departmental and state structures a way, and detail, determination of conformity of measured parameters to regulated magnitudes of maximum concentration limit and a maximum permissible emission.

Central tasks, which were put by working out of a new build of a rotoklon with inner circulation of a liquid:

- to produce criteria of a technical and economic estimation of a system effectiveness of environment protection against pollution;
- to create the apparatus with a wide range of change of operating conditions and a wide scope, including for clearing of gases of the basic industrial assemblies of a finely divided dust [9–13].

14.2   CURRENT STATE OF A PROBLEM

Optimization problem it is close, it is connected with various alternatives of designs, characteristic for designing a gas-cleaning installation of constructions. The survey of the domestic and foreign literature has revealed a total absence of attempts of the solution of a problem of optimization in the stated aspect. Moreover, it is not revealed attempts to optimize are constructive-technological parameters of any unique apparatus within the limits of its use in concrete conditions.

Preliminary studying of a problem has shown its complexity. The way to its solution as it is installed, passes through construction in the mathematical form of technical and economic models of optimized installations, i.e., the equations in which is constructive-technological and technical and economic parameters would be connected together. Now such model exists only with reference to cyclonic installations [14–20].

The special attention is given to the preparation and the analysis of technical projects on designing as shows the experiment bases of quality of the design that are put at the early stages of work.

It is possible to note the momentous defects to overcome, which it is necessary for the proximal years.

1. Insufficiency of the nomenclature a gas-cleaning installation and its lag from growing powers of the industry.
2. Weakness of computational baseline in which predominates empiric.
3. Absence of strict scientific criteria for designing a gas-cleaning installation of constructions with number of steps of clearing two and more. For the specified reason at designing of such constructions, the big role is played by purely heuristic factor.
4. The weakest and improbable working out of the questions connected with drawing by the flying emissions of a damage to a cir-

cumambient and, accordingly, with definition of economic benefit of liquidation of this damage.

These and other unresolved problems should be solved to the development engineers, initiating to master designing a gas-cleaning installation of constructions [21–24].

14.3   CALCULATION OF THE PREVENTED DAMAGE FROM ATMOSPHERIC AIR POLLUTION

The prevented ecological damage from pollutant emissions in an aerosphere—an estimation in the monetary form of possibly negative after effects from pollutant emissions, which is an observed time span, it was possible to avoid as a result of activity of supervising authorities in metrology, conducting of a complex of provisions, implementation of nature protection programs. At implementation a gas-cleaning installation was momentous to define magnitude of an economic damage in the set region, prevented as a result of conducting of nature protection provisions on air protection from pollutant emissions [25–34].

The integrated estimation of magnitude of the prevented damage from pollutant emissions can be spent to an aerosphere as for one large source or group of sources, and for region as a whole. In the capacity of sized up group of sources, all sources in the given city, the region, observed as a uniform source can be observed.

The prevented economic damage from pollutant emissions in a free air $_{Ur}$, thousand rbl/in observed economic region Russian Federation for a design time span is defined year by formula:

$$Y_p = Y_y \times M_{pr} \times K_e \times I_d$$

where Y_y—a parameter of a specific damage from pollutant emissions in a free air in observed economic region Russian Federation, Russian rouble/ton.

K_e—factor of an ecological situation and the ecological significance of a condition of a free air of territories of economic region of Russia.I_d—an index on the industries, installed by Ministry of Economics of Russia. We accept equal 1;

M_{pr}—a pollutant emission reduced mass in the region, scaled down as a result of conducting of matching nature protection provisions, 1,000 tons/year.

K_{ei}—factor of relative ekologo-economic hazard to ith pollutant.

N—quantity of considered pollutants.

$$M_{pr} = \Delta M - M_{cn}$$

$$\Delta M = M_1 - M_2 + M_{new}$$

where ΔM—total volume of a reduced mass of the scaled down waste interception, thousand conditional, tons/year;

M_c—a reduced mass of the waste interception, which has been scaled down as a result of slump in production in region, thousand conditional, tons/year;

M_1 and M_2—accordingly a waste interception reduced mass on the beginning and the end of the design period, thousand conditional, tons/year; M_n—a reduced mass of waste interception of the new factories and manufactures, thousand conditional, tons/year.

For calculation of a reduced mass of pollution, the confirmed values of maximum-permissible concentration (maximum concentration limit) of pollutants in water of ponds are used. By means of maximum concentration limit factors of ekologo and economic hazard of pollutants as magnitude, return maximum concentration limit are defined:

where $\mathbf{K}_{ei} = 1/M$

The reduced mass of pollutants pays off by formula:

$$M = \sum_{i=1}^{N} m_i \cdot K_{ei}$$

where m_i—mass of actual waste interception of pollutant in water installations of observed region, tons/year; K_{ei}—factor of relative ekologo-economic hazard to pollutant; N—quantity of considered pollutants. i—Substance number in the table.

14.4 TECHNICAL AND ECOLOGICAL ASSESSMENT GAS-CLEANING PLANT SAMPLING

In ecology and harmonious exploitation bases, estimations of economic efficiency of nature protection provisions are resulted. The problem is put to inject into calculation of a damage to a circumambient Y operational parameters of the given clearing installation to pass to relative magnitudes. It will allow to scale down number of the factors, which are not influencing functioning of system, to devise methods of calculation of relative efficiency a gas-cleaning installation of the constructions, giving the chance to choose the most rational approaches and the equipment of systems of trapping of harmful making atmospheric emissions. In the most general event, the damage A, caused by atmospheric emissions can be computed as $Y = B \cdot M$. The Reduced mass of emission incorporating N of components, will be computed in an aspect:

$$M = \sum_{i=1}^{N} A_i \cdot m_i$$

The emission mass m_i is proportional to an overshoot through system

$$m_i = (1 - \eta_i) \cdot m_{oi}$$

In an industrial practice, it is normally set or the share of concrete pollution in departing gas is known, C_{oi}. We will consider gas diluted enough so its density ρ does not depend on presence of admixtures

$$m_{oi} = C_{oi} \cdot \rho \cdot Q$$

Let us compute a damage caused to an aerosphere, per unit masses of trapped pollution Y_m

$$Y_m = \frac{B \times \sum_{i=1}^{N} A_i \times (1 - \eta_i) \times C_{oi}}{\sum_{i=1}^{N} \eta_i \times C_{oi}}. \tag{14.1}$$

If to size up the gas-cleaning plant on average parameters, $h_1 = \eta$

$$A = \frac{1}{N} \cdot \sum_{i=1}^{N} A_i \cdot C_{oi} \quad Y_m = \frac{B \cdot A \cdot (1 - \eta)}{\eta}$$

Let us formulate a principle of ecological efficiency of nature protection provisions at least a damage put to a circumambient. Purpose function in this case will appear in the form of $Y_m \rightarrow min$. Magnitude Y_m diminishes with value growth

$$E = \frac{\sum\limits_{i=1}^{N} \eta_i \times C_{oi}}{B \times \sum\limits_{i=1}^{N} A_i \times (1 - \eta_i) \times C_i} \tag{14.2}$$

Magnitude E, we will consider as criterion of ecological efficiency of nature protection provisions. The criterion of relative ecological efficiency Θ is representable in the form of the relation of values E computed for compared alternative E_1 and the base accepted in the capacity of E_0

$$\Theta = E_1 / E_0$$

In case of the single-component pollution of criterion of relative ecological efficiency we will find as

$$Q = \frac{h_1}{h_i} \times \frac{1 - h_i}{1 - h_1} \cdot \tag{14.3}$$

Thus, for two gas-cleaning plants of the concrete manufacture, different in separation extent, $h_i \neq \eta_0$, relative ecological efficiency of system is sized up by technological parameter $\Theta \rightarrow max$. The prevented damage Y_r compute as a difference between economic losses of two competing alternatives as $Y_p = Y_0 - Y_1$.

We will be restricted to an event of comparison of two alternatives of the clearing of gas emissions intended for the same manufacture with fixed level of technological perfection. In the capacity of base alternative Y_0, we will accept the greatest possible damage atmospheric emissions of the manufacture, which technological circuit design does not provide a clearing stage, $h_{10} = 0$. For the fixed technological circuit design of manufacture efficiency of a stage of clearing, we will size up in shares from the maximum damage $E_n = Y_p/Y_0$

$$E_n = \frac{Y_p}{Y_i} = \frac{\sum_{i=1}^{N} A_i \times C_{Hi} \times \eta_i}{\sum_{i=1}^{N} A_i \times C_{Hi}}. \tag{14.4}$$

To consider that all components of harmful emission with aggression average indexes are trapped in equal extents $(A_i = A,\ \eta_i = \eta)$, we come to $E_n = \eta$. Thus, the eurysynusic extent of trapping η is a special case of criterion of ecological efficiency E_n computed for one-parametric pollution or for emission with average characteristics. At sampling of the gas-cleaning plant, it is necessary to give preference to the installation securing higher values of criterion E_n.

The gas-cleaning installation demands expenses Z on the creation and functioning. These charges can essentially differ depending on the accepted method of clearing of gas emissions and should be taken into consideration at an estimation of the general damage. For example, air purification from a dust cyclone separators will be more low-cost in the "dry" way "wet" at which it is necessary to provide additional charges on water, pumping devices, neutralization of runoffs, etc. At the same time, centrifugal separators are not suitable for clearing of gaseous impurities. We will use relative parameters, i.e., to consider a gain of the prevented damage $DY_p = Y_1 - Y_2$ on rouble of expenses ΔZ.

Function of the purpose Y_p max will register in an aspect

$$E_n = (DY_n/DZ) \rightarrow \text{max}.$$

We will be restricted to consideration of a method of calculation of magnitude Y_p for centrifugal dust traps widely used in practice. We will define in maintenance costs a variable component of the power inputs connected with a water resistance of the apparatus DP (Pascal). The Head loss $\Delta H = DP / \rho$ (J/kg) is definable from the Bernoulli's theorem, which has been written down for entrance and target cross-sections of the gas passage. The energy consumption will be computed as $I = DP\ Q$(J/s).

Power inputs Z_e in terms of energy costs C_3 (Russian rouble/J) is definable from a relationship

$$Z_e = C_e \cdot Q \cdot \Delta P$$

The prevented damage Y_p computed on rouble of expenses, we will find as

$$\mathring{A}_i = \frac{\hat{A} \cdot \rho \cdot \sum_{i=1}^{N} A_i \cdot C_6 \cdot \eta_i}{C_{\mathring{y}} \cdot \Delta \eth}. \tag{14.5}$$

The criterion of relative ecological efficiency of the whirlwind apparatus $\Theta_n = E_{n1}/E_{n0}$, is computed on values E_n for two installations E_{n1} and E_{n0}, from which one is accepted for base E_{n0}. At transition to average magnitudes

$$E_n = \frac{B \cdot \rho \cdot \sum_{i=1}^{N} A_i \cdot C_{oi} \cdot \eta_i}{C_{\mathring{y}} \cdot \Delta P}. \tag{14.6}$$

Let us apply the results received earlier according to efficiency of clearing of gas emissions to the comparative analysis of dust extractors of centrifugal action by criterion Θ. As the base two-stage installation of type "cyclone C-6" is accepted. Results of the comparative analysis are introduced in Table 14.1. As shown in the introduced data, criterion relation in technical and ecological efficiency Θ reflects logic of process of dust separation—the above a device purification efficiency η, the magnitude Θ is more. In this case, instead of qualitative ascertaining of the fact the quantitative assessment of efficiency of the clearing of gas emissions is offered, allowing to define in what degree competing systems differ from each other.

TABLE 14.1 Comparative technical and economic indicators of offered and base dust-collecting plants

Nop/p	The indicator name	Unit measurements	Basic ware	hard-	New installation
1	Productivity	m³/s	6		6
2	Hydraulic resistance	Pascal	2,100		1,350
3	Factor of hydraulic resistance		10.8		6.9

TABLE 14.1   *(Continued)*

Nop/p	The indicator name	Unit measurements	Basic hardware	New installation
4	Concentration of emissions after installation	Mg/m³	127.5	39
5	The occupied space in the plan	m²	17	5.3
6	Metal consumption	m²	6.8	4.9
7	The general power consumption	kilowatt·h	30	31
8	The specific expense of the electric power on clearing 1,000 m³ emissions	kilowatt·h	0.6	0.475
9	Intensity of flow	kg/h	-	0.5
10	Criterion of efficiency θ		1.0	1.85

On Figure 14.1, results of researches of separating ability of a rotoklon depending on conditional speed of gas v_y, calculated on full section of the device are introduced. Comparison of considered devices by means of criterion relative technical and economic efficiency θ is spent. The cited data has illustrative character and shows possibilities of application of criteria of technical and economic efficiency θ for comparison purposes gas-cleaning plant devices.

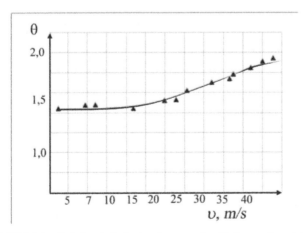

FIGURE 14.1   Relative technical and economic efficiency of a rotoklon.

14.5   RESULTS OF INDUSTRIAL TESTS OF THE GAS-CLEANING PLANT ON THE BASIS OF DEVICE "ROTOKLON"

Let us consider the system of wet clearing of the gases departing from the closed ferroalloy furnace 1. On this furnace, comparative researches of the described system of wet dust separation (see Figure 14.2) have been conducted.

The slope breeching 2 actually is the hollow scrubber in diameter of 400mm working in the evaporation cooling regime. At work gases arrive from it in a Venturi scrubber 3, consisting of two cylindrical columns in diameter of 1,000mmwith the general bunker. In each column of a scrubber, it is established on three atomizers. The Venturi scrubber of the first step of clearing has a mouth in diameter of 100mm and is irrigated with water from an atomizer established in front of the confusor.

Gases after a slope breeching go at first to the bunker the drop catcher 4, and then in a rotoklon of the another step, which consists from inertial a heat—and a mist eliminator 7. The exhaust of gases from the furnace is carried out by vacuum pump VVN-50 established behind devices of clearing of gas emissions. Purified gases are deduced in atmosphere.

Regulation of pressure of gases under a furnace roof and the expense of gases is carried out by a throttle in front of the vacuum pump. Slurry water from devices of clearing of gas emissions flows off a by gravity in a tank of a hydroshutter 9, whence also a by gravity arrives in a slurry tank. From a slurry tank water on two slurry clarifier is taken away on water purification. After clarification, chemical processing and cooling water is fed again by the pump on irrigating of gas-cleaning installations(Figure 14-2].

The dust containing in gases, differs high dispersion (to 80 weight % of particles less than 5–6mm). In Table 14.2, the compound of a dust of exhaust gases is resulted.

In tests for furnaces almost constant electric regime that secured with identity of conditions at which parameters of systems of dust separation characterize was supported. The furnace worked on the fifth—the seventh steps of pressure at fluctuations of capacity 14.5–17.5mW.

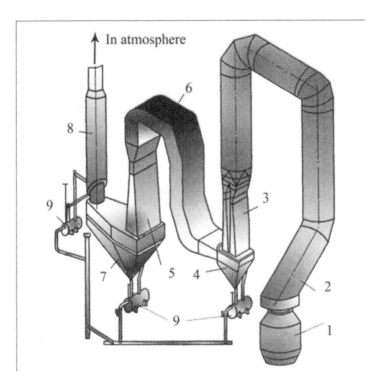

FIGURE 14.2 The scheme of clearing of flue gas with gas cooling in a venturiscrubber and the subsequent clearing in a rotoklon: (1)—baking oven; (2)—gas exit branch; (3)—venturi scrubber; (4)—bunker the drop catcher; (5)—arotoklon; (6)—gas pipeline; (7)—inertial heat and a mist eliminator; (8)—exhaust pipe; and(9)—tank a hydraulic hitch.

The quantity of dry gases departing from the furnace made 1,500–2,000m³/h. The temperature of gases before clearing of gas emissions equaled 750–850°C, and humidity did not exceed 4–5 percent (on volume).

TABLE 14.2 Results of post test examination

Compound	Requisite concentration (g/m³)	Concentration after clearing (g/m³)
Dust	0.02	0.00355
NO₂	0.10	0.024
SO₂	0.03	0.0005
CO	0.01	0.0019

TABLE 14.3 Results of calculation of a payment for pollutant emission

The list of pollutants (the substance name)	It is thrown out for the accounting period, t/year			The base specification of a payment within admissible specifications, a Russian rouble/t	The size of a payment for a maximum permissible emission, Russian rouble/year	The base specification of a payment within the established limits, a Russian rouble/t	Total a payment on the enterprise, a Russian rouble/year
	In total	including					
		VPE	MPE				
1	2	3	4	5	6	7	8
The inorganic dust	19 710		—	21	228.65	105	228.65
Nitrogen dioxide	105.1		—	52	379.19	260	379.19
Carbon monoxide	288.2		—	0.6	2.99	3	2.99
Sulfurs dioxide	197.1		—	40	539.14	200	539.14
Total:					1149.97		1149.97

In Table 14.3 results of calculation of a payment for pollutant emission of system of dust separation are shown:

Thus, we have chosen the scheme of clearing of gases which allows to lower concentration of pollutants to preset values and consequently, and to lower payments by the enterprise for emissions.

14.6 CONCLUSIONS

1. On the basis of a method of an estimation of economic efficiency of the carried out nature protection actions parities for calculation of the damage put to environment by atmospheric emissions of manufacture are received. Transition to relative indicators has allowed to scale down number of the factors, which are not influencing process of clearing of gas emissions

2. Methods of an estimation, a gas-cleaning installation of the constructions are devised, allowing on a design stage to make the comparative analysis of competing systems in terms of expenses for realization of nature protection actions

3. Criteria of an estimation technical and ecological, the gas-cleaning plants, incorporating both economic, and technology factors are developed.

KEYWORDS

- **Atmospheric emissions**
- **Criterion of efficiency**
- **Ecological efficiency**
- **Maximum permissible concentration**
- **The ecological damage**
- **The rotoklon**

REFERENCES

1.   Belevitsky, A. M.; Designing a Gas-Cleaning Installation of Constructions. L: Chemistry;**1990.**

2. Belevitsky, A. M.;Economy and Technical and Economic Optimisationof Dust-Collecting Plants (on an Example of Installations of Cyclonic Dust Separation): Help Supervising Material. Clearing of Gas Emissions L.;**1982.**

3. Berezhinsky, A. I.; and Homutinnikov, J. S.; Recycling, Cooling and Clearing of Gases. M: Metallurgy;**1967.**

4. Weinstein, M.; Krasnik, I. M.; and Durable, C. D.; Industrial and Sanitary Clearing of Gases. Tsintihimneftemash. M.;**1977,** 5.

5. Vilesov, N. G.; and Kostjukovsky, A. A.; Clearing of Gases. Kiev: Technics;**1971.**

6. A Time Typical Technique of Definition of Economic Efficiency of Realisationof Nature Protection Actions and an Estimation of the Damage Caused to a National Economy by Environmental Pollution. L.: Gidro-Meteo-Publishing House;**1986.**

7. The equipment for clearing of gas emissions. In: The Catalogue Environmental Control. Scientific Research Institute Gas Fine Gauge Strainers,M.;**1992.**

8. Denisenko, L. I.; et al.Quality Monitoring and Clearings of Plant Emissions of Various Manufactures of Fluoric Connections. M.: Tsintihim-Neftemash. **1982.**

9. Dubchak, R. C.; Waste Regaining of AluminiumManufacture Abroad. M.: Nonferrous Metals the Information;**1978.**

10. Zubenko, J. D.; and Ilyin, A. A.; Optimisation the Decision of Industrial Problems. M: Statistics;**1977.**

11. Ivanov, I. P.; Kogan, B. I.; and Bulls, A. P.; Engineering Ecology. Ed. Kogana, B. I.; Novosibirsk: NSGTU;**1995.**

12. Kazakevich, V. V.; System of Automatic Optimisation. M.: Energy;**1977.**

13. Smiths, A. E.; and Troitsk, T. M.; Protection of Air Pool against Pollution by Harmful Substances of the Chemical Enterprises. M.: Chemistry;**1979.**

14. Kouzov, I. A.; Malgin, A. D.; and Skryabin, M.; Dust Element of Gases and Air in a Chemical Industry. L.: Chemistry;**1982.**

15. A design procedure of concentration in atmosphere of the harmful substances containing in emissions of the enterprises. In: The Specification -86. M.: Stroyizdat;**1975.**

16. Nikanorov, A. A.;Sources and Methods of Clearing of Gas Bursts отOzone. M.: Nonferrous Metals the Information;**1985.**

17. The GeneralPrinciples of Technical Exploitation газоочистного and the Dust Removal Equipment at the Industrial Enterprises. M.: The Ministry of Chemical Mechanical Engineering of the USSR;**1971.**

18. Clearing and a Regeneration of Plant Emissions. Ed. Maksimova, V. F.;M.: The Wood Industry;**1981.**

19. Clearing of Technological Gases. Ed. Semenovoj, T. A.; and Lejtesa, I. L.;M.: Chemistry;**1977.**

20. EnvironmentalControl. Ed. Belova, S. V.;M.: The Higher School;**1991.**

21. "The governmental order of the Russian Federation No 632 from August 28, 11" About the statement of an order of definition of a payment and its limiting sizes for environmental pollution, housing of a waste, other kinds of damage effect (in ред. From June 14, 2011 with changes from May 14, 2009).

22. An EnvironmentProblem in Economic and the International Relations. *Collective of* Authors. M.:Thought;**1976.**

23. Rodions, A. I.; Klushin, C. H.; and Sister, C.; Technological Processes of Environmental Safety. Kaluga: N. Bochkarevoj's Publishing House;**2000.**

24. Straus, V.; Industrial clearing of gases. The Lane with English.M.: Chemistry; **1981.**
25. Skryabin, J. I.; The Industrial dust Atlas.M.: Tsintihim-Neftemash;**1982.**
26. Semenova, T. A.; et al. Clearing of Technological Gases. M.: Chemistry;**1969.**
27. Technics of environment protection. In: The Textbook for High Schools. Eds. Radionov, A.I.; Klushin, V. N.; and Gorshechnikov, N. S.; M.: Metallurgy;**1989.**
28. Feodors, E. T.;Society and Nature Interaction. L.: Gidro-Meteo-Publishing House;**1972.**
29. Tsygankov, A. P.; et al. Technical Progress-Chemistry-Environment. M.: Chemistry;**1979.**
30. Cyclones Scientific Research Institute Gas: Guidelines on Designing, Manufacturing, Installation and Exploitation. Yaroslavl: Seven-Bratovsky Branch Scientific Research Institute;**1971.**
31. Shvez, N. M.; Technical and Economic Researches of Schemes of Clearing of Gases from a Dust in Ferrous Metallurgy: The Express Information. Sulfurs,M.;**1967.**
32. EnterpriseEconomy: The Textbook for High Schools under the Editorship of Prof. Gorfinkelja,V. J.;Prof. Shvandara, V. A.;M.: It JUNITI-IS Given;**2007.**
33. EnterpriseEconomy: The Textbook for High Schools. Eds. Taburchaka, P. P.; and Tumina, V. M.; SPb: Chemistry; **2006.**
34. Shvez, N. M , Economy of Neutralisationof Gas Bursts. M.: Chemistry;**1979.**

PART III
BUBBLING-VORTEX APPARATUSES

CHAPTER 15

SURVEY OF ALTERNATIVES OF A DESIGN OF VORTEX GENERATORS

CONTENTS

15.1 INTRODUCTION

Multistage whirlwind apparatuses (WA) with the general counterflow motion of co-operating streams on a build are analogous to columns of bubbling type but has also some differences causing them of advantage (Frolov, 1987; Nikolaev, 1907; and Nikolaev, 1971). The first difference consists that on plates one or several rotary connections (RC) with direct-flow interacting of phases become stronger. Gas (steam) with a great speed is included into a RC, picks up a liquid arriving through special channels, and involves it in joint unidirectional, direct-flow motion. On an exit from the rotary connection, the stream of steam (gas) separates from a liquid[1–10].

At direct-flow interacting of streams, the liquid from a contact zone is taken out with a great speed. Tests have shown that the most effective branch of a liquid from gas (steam) is attained in the field of a centrifugal force. Presence whirlwind, is forward-rotary motion in a RC is another feature of whirlwind apparatuses[11–16].

The high speed of the easy phase, which are carrying away in direct-flow motion a liquid, and reliable separation of phases after interacting in a vortex flow allow to intensify sharply a mass transfer and in 3–8 times to increase carrying capacity of the apparatus. Application of whirlwind apparatuses for a dust separation (Ershov, 1973; Prihodko, 1979; and Diarov, 1979) is presented.

Known builds of direct-flow-vortex apparatuses are offered for classifying on set of characteristic signs (Nikolaev, 1970):

Characteristic attributes	Kinds, types
Quantity of steps	Single—stage
	Multi—stage
Kind of steps	Single-element
	Multielement
Direction of movement of streams	Descendingparallel flow
In a zone of contact	Ascendingparallel flow
	Cross parallel flow
Movement of a liquid through	Without circulation of a liquid
Contact zone	With partial recirculation
	With full recirculation
Type vortex generator	Axial
	Tangential

	Transitive
Way of submission of a liquid	Peripheral
In a zone of contact	Central
	Combined
Way of separation	Centrifugal—gravitational
	Centrifugal—inertial
	Centrifugal—filtration

In devices with descending parallel flow, a liquid and pairs (gas) in a zone of contact move from top to down, therefore they can work in very wide interval of loadings on a liquid and gas. But in this case for preservation of the general countercurrent in the device additional channels for a supply of gas (pair) on overlying steps are required. Therefore, such devices are recommended for processes with the greater specific charge of a liquid, and also for processes of gas purification with one step of contact (Mitropolskaja, 1974; and Uspenski, 1978).

In devices with cross parallel flow, there is a local unidirectional movement of streams, as a rule, to the subsequent centrifugal—gravitational separation. Speed limit of an easy phase in devices of this type makes 6–10mps. They are expedient for using, for example, in processes with significant thermal effects (Karpenkov, 1970; and Vazovkin, 1972).

Most high efficiency possess (WA) with ascending parallel flow. In a zone of contact of such devices speed of an easy phase makes of 15–40mps at average pressure. They are widely applied in the industry and are the most perspective[17–24].

15.2   DESIGNING OF ROTARY CONNECTIONS

Basic element of whirlwind apparatuses is whirlwind contact devices (WCD). They consist of a contact connecting pipe, an air swirler, knots of input of a liquid in a contact, and separation zone[25-26].

Motion in a WCD is progressive-rotary is created by the axial, tangential, or combined air swirlers. In a WCD of small diameter (20–50 mm) I, install dual-lead spirals (Figure 15.1) (Sergeys, 1972) or use a tangential-slot-hole air swirler (Figure 15.2) (Uollis, 1972). The spiral represents the metal strip twisted along a longitudinal axis. A slot-hole air swirler has milling of tangential slots in the most contact connecting pipe.

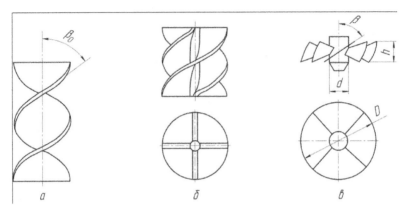

FIGURE 15.1   Axial air swirlers (a)–dual-lead spiral; (b)–the multiple-start screw insert; and (b–a)multibladeaxial air swirler.

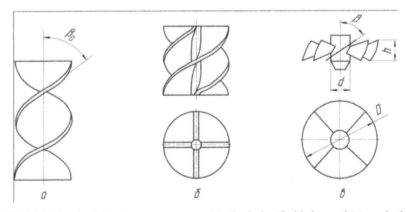

FIGURE 15.2   Combined vortex generator: (a)–slot-hole; (б)–blades; and (в)–cyclonic.

The greatest distribution have received multiblade axial vortex generator (Figure 15.2) (Abramovich, 1953) and tangential-blade (Figure 15.2, 6) (Tananajko, 1975), applied in (WCD) diameter 50–150mm. They consist from directing flat or profile, the blades forming channels for pass of gas (pair). The first differ the compactness, the second provide more intensive mass transfer.

Combined vortex generator (Figure15.3) (Kutateladse, 1976; Mamayev, 1978; Nurste, 1973; and Galkovski, 1979) posses the improved characteristics, but because of relative complexity are recommended for (WCD) the big diameter.

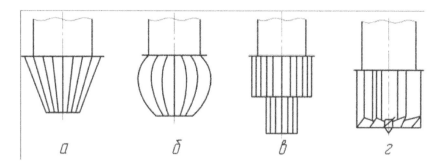

FIGURE 15.3   Combined vortex generator: (a)–conic; (б)–tangential-curvilinear; (в)–tangential-step; and (г)–tangential-axial.

Vortex generator is characterized by relative section, a degree of overlapping and factor twist. The relative section is an attitude of the total area of cross-section sections of gas channels to cross-section section of a contact branch pipe. The factor twist n is defined as the attitude of quantity of movement in a tangential direction to quantity of movement in an axial direction. The degree of overlapping S is the attitude of a projection of length of the gas channel on a plane vortex generator to width of the channel in this plane.

In Table15.1 parities on which the specified parameters pay off are resulted.

TABLE 15.1   Calculation of characteristics vortex generator

Type vortex generator	Relative sectionf	Factor twistn	Degree of overlapping S
Spiral			
Multibladeaxial			
Tangential Slot-hole			
Tangential Blade			

Notes: 1.z, δ- number and thickness blade accordingly; b- Width of the channel for pass of a liquid; ϒ- a corner between an axis of the channel and radius (cd); H- height vortex generator; h- width of the blade vortex generator; D-diameter (cd); d- diameter of the central plug.

2. For raise of efficiency of an eddy generation shutdown extent. It is recommended to choose within 0.2–0.3(Table 15.2).

TABLE 15.2   Definition of factor twist for (wd) with ascending parallel flow

Diameter of the contact device	Type vortex generator	Degree twist	Input of a liquid	Relative height of a contact branch pipe H/D	Type of a separator
20–40	Spiral or tangential-slot-hole	0.5–1.0	Peripheral above vortex generator	5–8	With slot
50–100	Axial blade	1–1.5	Central under vortex generator	2–5	Foramina's or with shutoff device liquids
100–200	Tangential blade	1–1.5	Central in vortex generator	1–2	Tangential-slot-hole or with shutoff device liquids

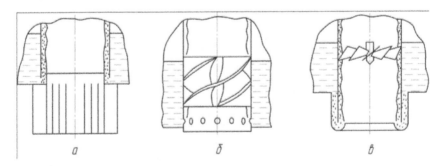

FIGURE 15.4   Peripheral input of a liquid:(a);(б–a)ring backlash; and(в)–apertures.

In such (wcd) mass transfer, it is carried out only through a surface of a film of the liquid current on a contact branch pipe(Figure 15.4). The attitude{relation} of this surface to the charge of an easy phase at the set speed is proportional to size H/D of a contact branch pipe with which increase efficiency mass transfer increases, but thus specific metal consumption (cd) raises{increases} also.

For provision of sufficient efficiency of a WCD in diameter more than 50°mm are expedient for inducting a liquid in core zone. Such feeding into can be carried out through the radial, axial, U- and G-shaped tubules also above or below an air swirler (Figure 15.5) (Nurste, 1973; and Kiselev,

1967). At liquid feeding into in a zone over an axial air swirler (Figure 15.5a) is formed a cone-shaped spray of liquid drops.

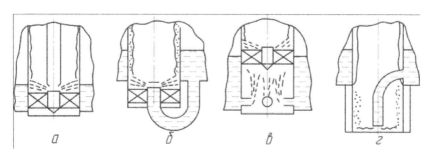

FIGURE 15.5   The central input of a liquid through branch pipes: (a)–axial; (б)–shaped; (в)–radial; (г)–shaped.

Gas penetrates this torch and intensively cooperates with a liquid. Under action of centrifugal force, the liquid is besieged on a wall of a contact branch pipe and moves after it{her} in the form of a film, continuing to contact to gas. In case of a supply of a liquid under axial vortex generator (Figure 15.5), all over again is formed an axial torch atomizer. At passage of a biphasic stream through vortex generator, there is a partial sedimentation and repeated dispersion liquids owing to what mass transfer it is in addition intensified.

In (vcd) with tangential-blade vortex generator, the liquid is entered into a ground part vortex generator (Tananajko, 1975; and Mamayev, 1978). Under influence, a gas stream the ring layer of drops, which collisions with edges blade and are in this zone of contact much longer, than in a torch atomizer (vcd) with axial vortex generator is formed. Keeping ability tangential vortex generator can be increased by giving to it{him} of the curvilinear form (see Figure 15.3b) or embedding{building in} in the top part washers. However, in comparison with axial tangential vortex generator are characterized by higher hydraulic resistance.

For a supply of a liquid in a zone of contact, it is offered to use hollow blades axial, conic, or tangential vortex generator [12002D19]. But, they are complex{difficult} in manufacturing and reduce relative section vortex generator that limits area of their application. Diameter and height of contact branch pipes essentially influence metal consumption of the device and efficiency mass transfer. For an intensification mass transfer in a number of designs (wcd), it is offered to put{render} on a surface of contact branch pipes ledges or the plates increasing turbulence of a stream (Galkovski, 1979; and Levdanskij, 1974), the design of the no rigid contact branch pipe, capable to vibrate (Diarov, 1979). Is offered also, however, it complicates manufacturing (wcd) and raises{increases} their hydraulic resistance a little.

For branch of a procontacted liquid from an easy{a light} phase use separators with slot, cracks, shutoff device, apertures. At the moderate loadings, simple separators with horizontal or vertical slot (Figure15.6a), which are mainly used in (wcd) small diameter (Uollis, 1972; and Schukin, 1970) are effective enough. The greatest distribution separators with shut-off device a film of a liquid (have received Figure 15.6b) (Kutateladse, 1976; and Korotkov, 1972).

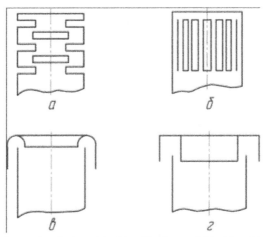

FIGURE 15.6   Separation devices (a), (b)–with slot; and (c), (d)–with shutoff device a film of a liquid.

And effective the foramina's separator with round apertures in which there is a gradual outflow of a liquid from a zone of contact is structurally simple. At very big loading use the combined separators having vertical slot or tangential cracks in a combination with shutoff device (Sergeys, 1972; and Levdanskij, 1974).

15.3   DESIGNING OF CONTACT ECHELONS OF WHIRLWIND APPARATUSES

Constructive registration of contact steps of industrial vortical devices is carried out in two basic directions.

The first direction assumes increase in diameter (wcd) with the central input of a liquid till the sizes of the device. Single-element steps are simple structurally, have low metal consumption. But radial non-uniformity of distribution of phases increases in them with increase in diameter, centrifugal force, and supposed axial speed of gas decrease. Therefore, in such devices parallel flow-cross interaction of phases with centrifugal–gravitational separation (Abramovich, 1953; and Nurste, 1973) is mainly used.

The second direction consists in creation of the contact steps consisting from several in parallel working (wcd) of small sizes. Much element contact steps provide uniform distribution of streams on section of the device that allows to create highly effective columns of the big productivity. The opportunity of detailed research hydrodynamical and mass transfer characteristics of separate elements in laboratory conditions causes reliability of design calculations. Vortical devices of various productivity have a high degree of unification that reduces expenses for their manufacturing. In (wcd) small diameter, the centrifugal force providing reliable separation at high loadings is significant. Therefore use of multielement contact steps is considered the most perspective in creation high-efficiency (wd) with parallel flow-ascending interaction of phases.

Multielement contact steps differ with design (wcd), and also the organization of movement of a liquid. The most simple are steps with recirculation liquids (Mamayea, 1978; Korotkov, 1972; and Levdanskij, 1974).

On them the liquid leaving a WCD, mixes up with its basic weight propelled on an echelon. Supply of a liquid from an echelon on an echelon is carried out on the general mine (Figure 15.7) or distributed tubular (Figure 15.7) to overflows.

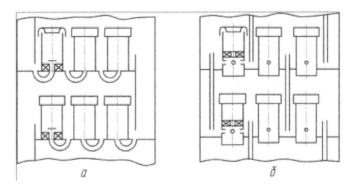

FIGURE 15.7   Contact steps with recirculation liquids: (a)–with the general{common} overflow; and (б)–with the distributed{allocated} overflow.

The first way allows increasing the length of "way" of a liquid by steps, and the second—enables more full to use section of the device.

Mixture of a procontacted liquid with initial causes reduction of motive power{forces} mass transfer process in this connection contact steps without recirculation liquids and with partial recirculation liquids have been developed. Full division of an initial and procontacted liquid can be reached{achieved} due to introduction of an additional horizontal partition (Figure15.8) (Nurste, 1973; and Korotkov, 1972) or uses individual flow channels (Figure 15.8b) (Mamayev, 1978; Galkovski, 1979; and Galeyev, 1976).

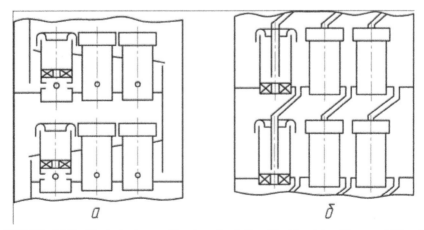

FIGURE 15.8   Contact steps without recirculation liquids: (a)–with an additional partition; and (b)–with individual overflow nipples.

Partial recirculation, it is created by installation on (wcd) below separators of inclined collars, reflective washers (Figure15.9) (Mamayev, 1978; and Nurste, 1973) or use of special system of liquid channels (Figure 15.9) (Galkovski, 1979). The raised{increased} metal consumption, complexity of manufacturing, and installation are the basic lacks of such steps.

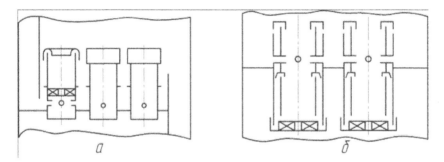

FIGURE 15.9 Contact steps with partial recirculation liquids: (a)–with collars; and (б)–with double input of a liquid.

Calculations show that negative influence of mixture at recirculation liquids can be essentially lowered by increase in time of stay of a liquid in a zone of contact. Therefore, the choice of a constructive variant of a contact step is defined{determined} by a parity{ratio} between the specific charge of liquid L/G for the device as a whole and optimum or limiting size of the specific charge ΔL/ΔG for (wcd) on the basis of comparative calculation.

At designing of multiple-unit echelons rotary connections on consecutive echelons, it is necessary to have coaxially and whenever possible on small distance from each other for conservation of a whirling motion of an easy phase (Sergeys, 1972; and Tananajko, 1975).

On contact echelons with recirculation of a liquid channels for liquid feeding into in a WCD, it is necessary to have so that to them got as less as possible a liquid which are getting out a WCD. The distance between echelons is defined either WCD altitude, or their water resistance and makes, as a rule, 300–600 mm.

15.4   CONCLUSIONS

The analysis of merits and demerits of various methods of clearing of gases from gaseous impurity has allowed to draw a conclusion that at clearing of great volumes of gas emissions by the most simple and in implementation the method of centrifugal separation is reliable.

However, traditionally applied separation equipment of packed (-column) and splash lubrication types possesses low carrying capacity on gas. In this connection, use of apparatuses of whirlwind type is in certain cases, the unique way of the solution of a problem.

Comparison of possibilities a gas-cleaning installation of various types also testifies to preference of application of whirlwind apparatuses at clearing of great volumes of gas emissions.

The analysis of known builds of rotary connections has allowed to draw a conclusion that the most preferable is use of a tangential air swirler which allows to vary extent of blade twist of a stream in wider limits

KEYWORDS

- **Blade twist extent**
- **Contact echelons**
- **Rotary connections**
- **Separation equipment**
- **The air swirler**
- **The irrigation water**

REFERENCES

1. Frolov, V. R.; and Ivanov, V. J.; High-speed mass transfer devices. The Survey Information. M.: "Energy";**1978.**
2. Nikolaev, N.A.;-Rectificative columns with vortical direct-flow steps.*TheoryBasis Chem. Technol.***1970**,*2,* 261–264,*(inRussian).*
3. Nikolaev, N. A.; and Short, J. F.; Mass transfer devices with vortical direct-flow steps. Newshigh schools of the USSR.*FoodTechnol.***1971**,*6,*135–138,*(inRussian).*
4. Ershov, A. I.; Plehov, I. M.; and Bershevic, A. I.; New of a design of separators for clearing industrial gases. The Survey Information. Minsk;**1973.**

5. Prihodko, V. P.; Safonov, E. N.; and Lebedyk, G. K.; Centrifugal drip pan with blade vortex generator. The Survey Information. Data Centre;**1979**.

6. Diarov, R. K.; Ovcinnikova, A. N.; and Nikolaev, N. A.; Device for separation and uniform distribution of multiphase streams on technological devices of preparation of oil. The Survey Information. Data Centre;**1979**.

7. Mitropolskaja, N. B.; Nikolaev, N. A.; and Bylkin, B. A.; Absorption the Device of High Efficiency for Complex Clearing Gases. News High Schools of the USSR, Chemistry and CemicalTechnology;**1974**, *1*, 151–153,*(inRussian)*.

8. Uspenski, V. A.; and Larin, J. K.; Research vortical turbulent gas washer. In: Collection "Industrial and Sanitary Clearing of Gases". Data Centre;**1978**, *6,3*–5,*(inRussian)*.

9. Karpenkov, A. F.; Development and research high-speed mass transfer the device of vortical type. Auto Abstract.Kazan: Institute of Chemical Technology;**1970**.

10. Vazovkin, E. S.; Research of hydrodynamics and efficiency of vortical contact steps. Auto Abstract. Kazan: Institute of Chemical Technology;**1972**.

11. Sergeys, A. D.; Research hydrodynamical regularity and mass transfer at ascending film current of a liquid. Auto Abstract. Kazan: Institute of Chemical Technology;**1972**.

12. Uollis, G.; One-Dimensional Biphasic Currents. "World";**1972**.

13. Abramovich, G. N.; Applied Gas Dynamics.Publishing House Technical Literatures;**1953**.

14. Tananajko, J. M.; and Voroncov, E. G.; Method of Calculation and Researches of Film Processes. Kiev, "Technique";**1975**.

15. Kutateladse, C. C.;and Stirikov, M. A.; Hydrodynamics Water Gas Systems. "Bowels";**1976**.

16. Mamayev, V. A.; Odisharija, G. E.; and Klapchuk, O. V.; Research of Process Mass Transfer. "Bowels";**1978**.

17. Nurste, N. O.; Cand. Dissertation Institute Heat-Electro Physicists.Tallin;**1973**.

18. Galkovski, V. J.; Research of laws of work of mass transfer vortical devices with tangential vortex generator. Auto Abstract. Minsk;**1979**.

19. Korotkov, J. F.; and Nikolaev, N. A.; Hydrodinamic of the Characteristic of Work Mass Transfer VorticalDeviceswith Tangential Vortex Generator. News High Schools of the USSR, Chemistry and Chemical Technology;**1972**, *5*, 800–804,*(inRussian)*.

20. Schukin, V. K.; Heat Exchange and Hydrodynamics of Internal Streams in Fields of Mass Forces. "Mechanical Engineering";**1970**.

21. Levdanskij, E. I.; Application of Contact Plates with Parallel Flow-Centrifugal Elements for an Intensification Mass Transfer Processes. Collection Data Centre;**1974**,*6*, 69–74,*(inRussian)*.

22. Galeyev, N. A.; Mustashkin, F. A.; and Nikolaev, N. A.; Hydrodynamical of law in contact devices mass transfer devices of vortical type. In: Collection: Machines and Devices of Chemical Technology. Kazan;**1976**. *4*, 42–44,*(inRussian)*.

23. Sabitov, S. S.; Issledovanie mass transfer in devices of parallel flow-vortical type. Auto Abstract.Kazan:Instituteof Chemical Technology;**1979**.

24. Karpenkov, A. F.; Nikolaev, H. A.;and Nikolaev, A. M.; Influence of a corner of an inclination blade vortex generator for work vortical mass transfer the device. Kazan: Collection Postgraduate Works, Institute of Chemical Technology;**1970**,*1*, 48–52,*(inRussian)*.

25. Kiselev, V. M.; Hydrodynamic of the characteristic and mass transfer on a cyclonic plate at adsorption dioxides of carbon. "Chemistry";**1967,** 7, 1630–1634,*(inRussian).*
26. Karpenkov, A. F.; Nikolaev, H. A.;and Nikolaev, A. M.; About an opportunity of increase of productivity mass transfer devices due to increase in speed of an easy phase. News High Schools of the USSR, Chemistry and Chemical Technology.**1971,***2,* 309–312, *(inRussian).*

CHAPTER 16

AERODYNAMICS OF VORTEX APPARATUSES

CONTENTS

16.1   INTRODUCTION

The formation of the turbulent twirled stream essentially differs from the forward. Under the influence of a centrifugal force in the twirled stream, there are pressure gradient on radius, the return currents, the raised speeds at a wall, non-linearity of a profile of tangential stresses, etc. In turn, these phenomena make strong impact on regularity of motion of drops of a liquid in a deduster and character of a current in a boundary layer on a phase boundary gas—a liquid, i.e., define both separation efficiency, and a water resistance, and a criticality of work of dedusters [1–9].

On character of distribution of axial speed, the twirled streams are classified as follows Frolov (1978):

(i)   Poorly twirled stream, which is characterized by that any cross-section, the axial component of speed has the maximum value on an axis. The profile of an axial velocity differs from a profile of a direct-flow stream a little.

(ii) Moderately twirled stream is characterized "undershooting" of axial making speed on a heading to an axis on a current. The reverse flow on an axis is not present; the axial velocity profile has the M-shaped form.

(iii) Sharp twirled stream is characterized by presence of a zone of return currents.

The offered classification of the twirled streams is not settling.

So the air swirlers, which design features strain velocity profiles, for example, are known in such a manner that, despite a considerable initial twisting of a stream, in axial area, there is no reverse flow or there are rather weak return currents. Besides, at small blade twists chances of reception of undershooting of axial making speed [10–17].

16.2   REGULARITY OF MOTION OF GAS AND A LIQUID

Researches show (Kutepov, 1977; and Uollis, 1972) that velocity profiles of a gas stream for geometrically similar air swirlers automodeling to diameter the rotary connection and to a gas rate. In the installed stream gate out an axial zone of quasifirm twirl and a peripheral zone of potential twirl with the matching approached profiles (Kutepov, 1977):

$$W_\phi = W_\phi^m \frac{r}{r_b}; \quad W_z = W_z^m \frac{r}{r_b} \quad (r \le r_b)$$

$$W_\phi = W_\phi^m \frac{r_b}{r}; \quad W_z = W_z^m \quad (r \ge r_b) \tag{16.1}$$

where W_ϕ and W_z—the maximum values of the peripheral (tangential) and axial speeds of gas accordingly;

r_b—radius of a zone of quasifirm twirl.

Directly over an axial air swirler where the vortex flow is formed, it is offered to present velocity profiles the approached equations (Uollis, 1972):

$$\frac{W_\phi}{\overline{\overline{W}}_z} = \frac{6}{f} \frac{(1-\overline{r})(\overline{r}-\varepsilon)}{(1-\varepsilon)^2} \sin\beta$$

$$\frac{W_z}{\overline{\overline{W}}_z} = \frac{6}{f} \frac{(1-\overline{r})(\overline{r}-\varepsilon)}{(1-\varepsilon)^2} \cos\beta \tag{16.2}$$

where $\hat{W}_z$—average speed;

$\check{r} = 2r/D$—relative radius;

$\varepsilon = d/D$—relative diameter of the central spigot.

Peripheral velocity presence causes emersion of the radial pressure gradient:

$$\frac{\partial P}{\partial r} = \frac{\rho_g \overline{W}_\phi^2}{r} \tag{16.3}$$

where ρ_g—gas density.

Using the Eq. (16.1) and considering that $f = cos\beta$, it is possible to find a difference of pressures ΔP

Between peripheral ($\check{r} = 1$) and axial ($r = 0$) zones:

$$\Delta P_r = 6n^2 \frac{\rho_g \overline{W}_z^2}{2} \tag{16.4}$$

Under the influence of the radial pressure gradient at an air swirler was possibly formation of a zone of a reverse flow (Pravdin, 1973). The liquid with peripheral supply moves to a RC in the form of a film on a spiral line, the average thickness of this film α_0 is on the equation (Guhman, 1986):

$$a_o = 16,46 \cdot 10^{-3} \left(\frac{q}{\cos \beta} \right)^{0,3} ; \quad \left(\frac{f}{\overline{W}_z} \right)^{0,75} \mu_l^{0,23} \quad \text{при} \frac{q}{\cos \beta} \leq 1,1$$

$$a_o = 16,6 \cdot 10^{-3} \left(\frac{q}{\cos \beta} \right)^{0,4} ; \quad \left(\frac{f}{\overline{W}_z} \right)^{0,75} \mu^{0,23} \quad \text{при} \frac{q}{\cos \beta} \geq 1,1 \qquad (16.5)$$

where μ—dynamic viscosity of a liquid, N/sm^2;

q—A water concentration, m^3/mh.

In the literature particularizes on hydrodynamics of film current (Karpenkov, 1960; Margolin, 1977; and Bunkin, 1970) are resulted {brought}. At input of a liquid in central part (cd), it {she} is splitter up for drops, which d_c average diameter is defined on the equation (Margolin, 1977):

$$\frac{d_m}{D} = 0,05 \frac{\sigma_l f^2}{\left(\rho_g \overline{W}_z^2 D \right)}^{0,35} \qquad (16.6)$$

where σ—a superficial tension.

Under action of centrifugal forces of a drop move to periphery and form on a wall of the contact cylinder a film, which parameters are defined from the Eq. (16.5).

The Height on which drops rise, pays off on a parity,

$$\Delta z = 0,25D / n \qquad (16.7)$$

Time of stay of drops in a vortical stream depends on radial speed v_r change can be found on the approached dependence:

$$\left(\frac{v_r}{\overline{W}_z} \right)^2 \approx 3\xi \frac{\rho_g}{\rho_l d_m} \cdot n\sqrt{1+n^2 r} \qquad (16.8)$$

where ρ—density of a liquid;

ξ—Factor of resistance of a drop.

16.3 FACTORS INFLUENCING AERODYNAMIC FORMATION OF A STREAM

For the characteristic of intensity of blade twist of an air stream created by various air swirlers, usually use a mean of speed of the stream, found on geometrical characteristics of apparatuses.

Strict enough definition of twisting ability of an air swirler on its geometrical elements is given in-process G. N. Abramovich (1953). Offered for calculation of centrifugal injectors the geometrical dimensionless characteristic will be defined as:

$$A = \frac{l \times W_\tau}{R \times W_a} \tag{16.9}$$

where W_t и W_a—averages under the charge of value of tangential and axial components of speed.

Parameter A sizes up a twisting in devices with a chamber air swirler when air is inducted into the chamber is strictly tangential. This formula, no less than others offered before the formula, it is possible to lead to an aspect,

$$n = \frac{M_c \cdot C}{K_c \cdot D} \tag{16.10}$$

where C—an arithmetical multiplier, equal $8/\pi$, injected for simplification of final formulas; D—diameter of the cylindrical channel.

$$M_c = 4\rho \cdot W_a \cdot W_\tau \cdot R^2 \cdot l, \quad K_c = \pi \rho W_a^2 R^2 \tag{16.11}$$

For reception of the design data characterizing intensity of blade twist, in-process (Abramovich, 1953) it is offered to operate with a mean axial W_a and tangential W_t components of speed of a stream, which can be easy are defined on the general air consumption.

Design data of intensity of blade twist n, inferred for various types of air swirlers in terms of approximating assumptions, are presented to Table 16.1. The parameter n on the physical sense is relative value of an entrance angular momentum.

Equality of design data n testifies to identity of aerodynamic characteristics of the twirled streams on an exit from geometrically similar air swirlers of the same type.

It is necessary to mean, however, that the real blade twist of a stream expressed in parameter,

$$\theta = \frac{M}{K \cdot R} \qquad (16.12)$$

where M—a stream angular momentum; K—a stream momentum; R—radius on an exit from an air swirler.

TABLE 16.1 Design data of air swirlers and aerodynamic characteristics of the twirled stream created by them

Type vortex generator	Symbols	Design data and mean regime characteristics
AV	l	$\dfrac{1}{3} \cdot \dfrac{d_2^3 - d_1^3}{d_2^2 - d_1^2}$
	$\dfrac{W_\tau}{W_a}$	$\dfrac{d_2^2}{d_2^2 - d_1^2} \cdot ctg\beta_2$
	n	$\dfrac{8d_2}{3\pi} \cdot \dfrac{d_2^3 - d_1^3}{\left(d_2^2 - d_1^2\right)^2} ctg\beta_2$
ATVP	l	$\left[r_1^1 + \dfrac{\left(d_1^| + 2d_1\right)\left(d_1 - d_1^|\right)}{6\left(d_1^| + d_1\right)} \right] \cos\beta_2$
	$\dfrac{W_\tau}{W_a}$	$\dfrac{d_1^2}{2h_1\left(d_1^| + d_1\right)\sin\left(\beta_2 + \dfrac{\pi}{z}\right)}$
	n	$\dfrac{8d_1\left(d_1^2 - d_1 d_1^| + d_1^{|2}\right)\cos\beta_2}{3\pi\left(d_1 + d_1^|\right)^2 \sqrt{4h_1^2 - \left(d_1 - d_1^|\right)^2}\sin\left(\beta_2 + \dfrac{\pi}{z}\right)}$

ATVC	l	$$\dfrac{\left[r_2^1 + \dfrac{\left(d_2^1 + d_2\right)\left(d_2 - d_2^1\right)}{6\left(d_2^1 + d_2\right)}\right]\cos\beta_2}{d_2^2}$$
	$\dfrac{W_\tau}{W_a}$	$$\dfrac{d_2^2}{2h_1\left(d_2^1 + d_2\right)\sin\left(\beta_2 + \dfrac{\pi}{z}\right)}$$
	n	$$\dfrac{8d_1\left(d_2^2 - d_2 d_2^1 + d_2^2\right)\cos\beta_2}{3\pi\left(d_2 + d_2^1\right)^2\sqrt{4h_1^2 - \left(d_2 - d_2^1\right)^2}\,\sin\left(\beta_2 + \dfrac{\pi}{z}\right)}$$
TBVP	l	$0,5d_1 \cdot \cos\beta_2$
	$\dfrac{W_\tau}{W_a}$	$\dfrac{\pi d_1^2}{4h_1 \cdot z \cdot \varepsilon}$
	n	$\dfrac{d_1^2 \cdot \cos\beta_2}{h_1 \cdot z \cdot \varepsilon}$
TBVL	l	$0,5d_2 \cdot \cos\beta_2$
	$\dfrac{W_\tau}{W_a}$	$\dfrac{\pi d_2^2}{4h_1 \cdot z \cdot \varepsilon}$
	n	$\dfrac{d_2^2 \cdot \cos\beta_2}{h_1 \cdot z \cdot \varepsilon}$

At the same value n cannot be equal to various types of air swirlers for some reasons, depending on the form of the setting of an air swirler. Major of them are:

(i) An air stream energy loss on an internal friction and a friction about walls of channels;

(ii) Irregularity of high-speed fields in various cross-sections of the setting of an air swirler;

(iii) Various conditions of formation of aerodynamic formation of a stream on an exit from an air swirler.

It is experimentally installed (Frolov, 1978) that values of parameters n and θ for air swirlers of type AV and ATVP not sharp differ among themselves. Real blade twist of a stream for air swirlers of type TBVP and

TBVL depends not only on design data n, but also from an outlet angle of a stream from an air swirler β_2.

Except design data n, the twirled stream in a deduster is characterized:

(a) In extent of irregularity of a velocity distribution.

$$\psi = \frac{W_{max} - W_{min}}{W_{cp}}, \% \tag{16.13}$$

(b) Magnitude of a zone of return currents (boundary line of this zone the surface on which axial speed is equal to null is);

(c) A stream angle of lead on a spiral β, which gives the indirect characteristic of extent of a twisting of a stream;

(d) A relationship of axial and tangential components of speed in a wall layer on a phase boundary (a gas-liquid film), defining a criticality of work of a separator.

Experiences show that the slope of guide vanes in air swirlers of type TBVP and ATVP makes strong impact on stream formation. The general for velocity profiles in air swirlers of a various aspect is that with increase in intensity of blade twist values of tangential components of speed increase.

The increase in a slope of guide vanes β_2 or decrease of an angle of conicity of air swirler ATVC results thereto that the maximum of speeds at enough big intensity of blade twist moves more close to a spinning ax.

This fact is remarkable that allows sampling of a matching slope of guide vanes and intensity of blade twist to have almost any sizes of a zone of recirculation up to its full elimination. This results from the fact that in air swirlers of type TBVP and ATVP, the gas make is carried out is not strictly tangential, and under some angle α. Thus, it is possible to have even such formation of a stream at which axial speeds in an axial zone will have the maximum positive values exceeding average speed more than three times.

At angle increase Υ transition from air swirler ATV to air swirler AV occurs. Thus, the zone of the maximum speeds moves to periphery. Tangential making speeds at angles $\beta_2 = 20°$ and $30°$ considerably exceed axial, accepting at coal $45°$ approximately equal values.

Within change of an angle β_2 from $30°$ to $45°$ return currents in an axial zone of a stream are observed. The zone of return streams at $\beta_2 = 30°$ attains $(0.5 - 0.6)d$.

The stream formation on small removal $x = 0.2 + 0.3$ from air swirlers of type TBVL at all values of a slope of guide vanes β_2 is characterized by presence of a powerful zone of return currents, almost equal to diameter of the overhead disk of an air swirler. The gas stream moves in the ring channel between the body and an air swirler, zones of the maximum axial, and tangential speeds here are had. Decrease of a slope of guide vanes by an exit from an air swirler β_2 leads to considerable decrease in axial speed in a peripheral zone. The maximum positive values of tangential speed thus can accept the values exceeding average speed more than four times.

In process of removal from an air swirler, there is a disproportionation of axial speeds on channel cross-section, and on distance $x = 0.8 + 1.0$ zone of return currents disappears.

Tangential speeds are redistributed more slowly; on considerable removal from an air swirler $x \geq 3D$, they continue to keep great values $W_t = 4 - 4.5$.

The analysis of the saved up experimental data (Abramovich, 1953) shows that presence of the central pipe does not influence essentially stream aerodynamic characteristics at its relative sizes, which are not exceeding halfs of diameter of the cylindrical channel $d_0/D \leq 0.5$. At $d_0/D \geq 0.5$, the aerodynamic formation of a stream undergoes already serious changes leading more to a uniform distribution of speeds in ring cross-section of the body.

Strong effect on aerodynamic formation of a stream leads to decrease of optimum diameter of the discharge connection d_3/D. A stream monkey wall, raising a water resistance on a way of its motion, promotes leveling of a high-speed field, and sharp abbreviation of a zone of negative streams. Performance of the discharge connection in the form of the conic diffusor leads to expansion of a zone of return current, however, irregularity of a velocity distribution in the discharge connection not only does not raise, but, on the contrary, decreases a little. The breakaway of a stream from diffusor walls in the tested range of angles of disclosing $2\,\alpha \leq 85°$ occurs only at low values of intensity of blade twist $n \leq 0.35$.

For a two-phase stream essential effect on aerodynamic formation of a stream magnitude of relative loading on a liquid $\acute{L} = L/G$ renders, kilograms per kilogram. Thus $\acute{L}$ practically does not influence distribution axial, but considerably reduces an average value of tangential speeds. The form of profiles $W_t = f(R)$ thus remains identical (Mustashkin, 1970; and Kutepov, 1977).

16.4 CONCLUSIONS

Design data of intensity of blade twist n, inferred for various types of air swirlers are observed. The parameter n on the physical sense is relative value of an entrance angular momentum. For the characteristic of intensity of blade twist of an air stream created by various air swirlers, it is offered to use a mean of speed of the stream, defined on geometrical characteristics of the apparatus.

The analysis of known builds of rotary connections has allowed to reveal major factors (loading on a liquid phase, a monkey wall, and a stream breakaway, discharge connection sizes), working upon aerodynamic formation of a stream.

All observed phenomena make strong impact on regularity of motion of gas and a liquid in a deduster, i.e., define both separation efficiency, and a water resistance, and a criticality of work of dedusters.

KEYWORDS

- **A velocity profile**
- **Aerodynamics**
- **Blade twist extent**
- **Intensity of blade twist**
- **Stream formation**
- **The momentum**

REFERENCES

1. Frolov, V. R.; and Ivanov, V. J.; High-speed mass transfer devices. The Survey Information, M.: "Energy." **1978.**
2. Ahmedov, R. B; et al. Aerodynamics of the twirled stream. M.: "Energy." **1977.**
3. Abramovich, G. N.; Applied Fluid Dynamics. M.: Publishing House the Tehniko-Theoretical Literature; **1953.**
4. Pravdin, V. A.; Aerohydrodynamic streams. *Chem. Eng. Ind.* **1973,** *10,* 35–37, (in Russian).
5. Mustashkin, F. A; Two-Phase Streams. Kazan: Works KCTU; **1970,** *45,* 94–99, (in Russian).

6. Karpenkov, A. F.; The Dust Separation Theory in the Turbulent Wet Washer. Works KCTU; **1970,** *45,* 40–45, (in Russian).
7. Nurste, H. O.; Ref. Diss. Cand. Tech. Sci. Institute Warmly—to Electrophysicists an Academy of Sciences of Est. Tallinn: The Soviet Socialist Republic; **1973.**
8. Schukin, V. K.; Heat exchange and hydrodynamics of internal streams in fields of mass forces. M.: "Engineering Industry." **1970.**
9. Pustotaja, V. F.; In The Collector: "Questions of Designing and Installation of Sanitary Systems." **1972,** *32,* 22–26, (in Russian).
10. Barahtenko, G. M.; In the Collector: "Cars and Apparatuses of Food Manufactures." Sverdlovsk; **1971,** *1,* 53–62, (in Russian).
11. Ershov, E. I.; et al. New builds of separators for clearing of industrial gases. The Survey Information. **1973.**
12. Margolin, E. V.; and Prihodko, V. I.; Perfection of production engineering of wet clearing of gas emissions at aluminium factories. M.: "Engineering Industry." **1977.**
13. Guhman, L. M.; and Minakov, V. V.; Cars and the oil equipment. M.: Higher School. **1986,** *9,* 20–22, (in Russian).
14. Bunkin, V. A.; Working Out and Research of the Mass-Transfer Apparatus with Direct-Flow Rotary Connections. Ref. Diss. Cand. Tech. Sci. Kazan: KCTU; **1970.**
15. Samsonov, V. T.; In the Collector: "Ventilation and an Air Conditioning at the Polygraphic Factories." Works Scientific Research Institute a Polygraph, M.: "Book." **1974.**
16. Kutepov, A. M.; et al. Hydrodynamics and heat exchange at steam formation. M.: "Higher school." **1977.**
17. Uollis, G.; One-dimensional two-phase currents. M.: "World." **1972.** Chuitt, J. G.; and Hall-Tejlor, H.; Ring two-phase currents. M.: "Energy." **1974.**

CHAPTER 17

EFFECT REGIME-DESIGN DATA ON EFFICIENCY OF CLEARING OF GAS AND A HYDRAULIC RESISTANCE

CONTENTS

17.1   INTRODUCTION

The majority of known methods of calculation and designing of whirl-wind apparatuses for conducting of processes of separation are based on methods of laboratory modeling and similarity theory. Such approach considerably narrows possibilities of designers at sampling constructive and operating conditions of the gas-cleaning installations intended for the solution of this or that technological problem, demands the big experimental works, occupies a lot of time, and not always can secure necessary level of accuracy of calculations.

Formation of streams in whirlwind apparatuses, their parameters of work (efficiency of conducting warmly—and mass-transfer processes, a water resistance, etc.) are defined by a build and sizes of whirlwind apparatuses and regimes of their work.

The methods of calculation, which are based on mathematical model, essentially speed up process of its working out and designing, sampling of a rational build of the apparatus intended for the solution of the set technological problem.

In any whirlwind apparatuses of a pressure loss are one of the major characteristics. They are necessary for knowing for sampling of fan devices, an estimation of efficiency of expenses of energy, comparison of various ways of conducting warmly- and mass-transfer processes, designing of apparatuses, their elements and systems.

Pressure losses in whirlwind apparatuses develop of following components:

(i)   Losses in upstream ends;

(ii)   Losses in air swirlers;

(iii) A friction loss in separation and bunker parts of the apparatus;

(iv) Losses at a gas entry in an exhaust connecting pipe and in it;

(v)   An additional pressure loss on a friction in a gas pipe, originating owing to blade twist of a stream, or the kinetic energy of gas connected with loss if it gets out an exhaust connecting pipe at great volume or an aerosphere (so-called noncondensing operation).

Most full the theory of a water resistance of whirlwind apparatuses is devised in Ref. Anthony (1970) and Uzhov (1973).

The water resistance is defined by the sum of pressure losses of an air swirler, in the separation chamber and on an exit from it. Thus, the cores are losses in an air swirler and on an exit. It proves to be true the data of tests of whirlwind apparatuses (Adamar, 1970; Jofinov, 1975; and Uzhov, 1973) with a different relative height of the separation chamber, which real values do not fall outside the limits 1.5–3.0.

Pressure losses in an air swirler depend on its geometry and extent of a twisting of gas. The exit loss depends on extent of a twisting of gas in the separation chamber and monkey wall extent on an exit, i.e., from the relation of diameters of the exhaust tube and the separation chamber.

Dependence $\zeta_{_2}$ and $\zeta_{_2}$ from type of air swirlers, blade twist parameters. To extents of a monkey wall of a stream on an entry has been studied (Karpenkov, 1970; Fucs, 1955; Bunkin, 1970; Ribchinskij, 1970; and Kutepov, 1976) at test of a great number of various constructive alternatives of whirlwind apparatuses. In the capacity of air swirlers used aksialno-VANED, tangential scroll air swirlers with direct and inclined upstream ends.

17.2   WATER RESISTANCE OF ROTARY CONNECTIONS

Hydraulic resistance usually represent in the form of the sum or product of two components,

$$\Delta P = \Delta P_{dry} + \Delta P_{g+l}$$

where ΔP_{dry} – hydraulic resistance of the dry device;
ΔP_{g+l} – hydraulic resistance water gas a layer.

Hydraulic resistance of the dry device characterizes losses of energy on sudden change of a direction and a configuration of borders of a stream, loss of energy on friction about walls of the case, and loss of energy in a target branch pipe.

The First kind of losses depending on change of the cross-section area of the channel and a direction of a stream in vortex generator, usually consider{examine} as local resistance, and losses on friction for technically smooth pipes in a wide interval of change of relative length of the case drip pan $\hat{H} \leq 100$ are insignificant (Samsonov, 1974). Generally, hydraulic resistance of the dry device represent in the form of the sum of two composed,

$$\Delta P_{dry} = \Delta P_{v} + \Delta P_{out} = \frac{\rho W_{cp}^{2}}{2}\left(\zeta_{v} + \zeta_{out}\right) \tag{17.1}$$

Under ΔP_{v} usually understand the sum of losses actually in vortex generator and the cylindrical channel behind it {him}, and ΔP_{out} considers reduction or increase in losses due to compression a stream on an output{exit} from the channel with its{his} subsequent expansion or narrowing (Kolborn, 1964; Kepelman, 1960).

Hydraulic resistance water gas a layer is defined {determined} basically by quantity {amount} of a liquid acting on separation (L, kg/kg) and way of its {her} tap {removal}. Absence of influence of physical and chemical properties of a liquid on hydraulic resistance drip pan testifies that power expenses for crushing of a liquid are insignificant in comparison with the general {common} expenses of energy (Kutepov, 1977).

Below some methods of calculation of hydraulic resistance of centrifugal devices with blade vortex generator are considered {examined}.

As hydraulic resistance of such devices is substantially defined {determined} by design features vortex generator, it is expedient to consider {examine} questions of hydraulic resistance, having taken for a basis the classification accepted earlier.

17.3   CENTRIFUGAL DEDUSTERS WITH AXIAL AIR SWIRLERS OF TYPE AV

From the numerous dependences offered by various authors (Margolin, 1977; Samsonov, 1974; Kutepov, 1977; Lang, 1964; and Marchall, 1977) for calculation of hydraulic resistance ΔP_v, the greatest interest is represented by in what influence of geometrical parameters vortex generator is considered.

In work (Lang, 1964) hydraulic resistance of the device with screw vortex generator is offered to be counted on the known equation Darsi modified with reference to a screw stream:

$$\Delta P_v = \frac{\rho W_{cp}^2}{2} \cdot \frac{h_1}{d_{ecv} \cdot \sin^3 \beta_2}, \qquad (17.2)$$

where h_1 – height vortex generator, m;

W_{cp} – Axial mean speed in the ring channel, mps;

$d_{ekv} = 4F/\Pi$ – Equivalent diameter of the screw channel, m;

F – The area of free section of the screw channel;

Π – The perimeter equal to the sum of the parties {sides} of a trapeze, m;

β_2 – The corner of an inclination screw directing to the horizon, taken on average diameter of a ring backlash, is glad.

In work (Margolin, 1977) in which results of researches screw vortex generator in a wide range of measurement of height of the screw are presented, it is shown, that dependence,

$$\Delta P = f\left(\overline{h}_1\right) \quad \overline{h}_1 = \frac{h_1}{l} \quad l = \frac{2S}{Z}$$

where h_1 – relative height vortex generator; l – the minimal height vortex generator at which overlapping {blocking} an axial stream by blades is reached {achieved}) has a minimum at $\hat{h} = 2$, and at $\hat{h}_1 \geq 2$ hydraulic resistance vortex generator practically does not change. According to the received experimental data, the optimum height vortex generator is offered to be accepted equal $\hat{h} = (2–2.5)l$, and hydraulic resistance to count under the formula,

$$\Delta P_v = \frac{\rho W_{cp}^2}{2} \cdot \frac{\left(r_2^2 + r_1^2\right)\left[\pi\left(r_2^2 - r_3^2\right)\right]^2}{16\left[\pi\left(r_2^2 - r_1^2\right) - z\delta\left(r_2 - r_1\right)\right]\sin^2\beta} \tag{17.3}$$

where r_3 – radius of the central pipe, m;
δ – Thickness of the blade, m;
W_{cp} – mean speed in separation to a zone, mps.
Data of works (Marchall, 1977; and Bralove, 1981) on hydraulic resistance in pipes with tape vortex generator are generalized by Schukin in work (Kutepov, 1977) in the form of dependence,

$$\zeta\left(\frac{D}{d}\right)^n = f\left[Re,\left(\frac{d}{D}\right)^2\right] \tag{17.4}$$

where D – diameter of curvature of an axial line of the channel, m.

$$\frac{D}{d} = 0,5 + \frac{8}{\pi^2}\left(\frac{S}{d}\right)^2 \tag{17.5}$$

where S – a step of a tape;
d – Diameter of a pipe, m.
At a turbulent mode of current for $S/D = 2.65 – 1.3$, the formula is received,

$$\zeta = \frac{0,705}{Re^{0,28}}\left(\frac{d}{D}\right)^{0,09} + 0,009\left(\frac{d}{D}\right)^{0,65} \tag{17.6}$$

It {She} is fair at $Re \leq 5.9 \cdot 10^4$. In the basic range of change $S/D \leq 1.75$. The deviation {rejection} of skilled data from settlement does not exceed 5 per cent.

At $S/D = 1.8 - 2.5$, dependence is received,

$$\zeta = \frac{4,72}{Re^{0,35}} \left(\frac{d}{D}\right)^{0,74} \tag{17.7}$$

This formula is fair at $Re(S/D)^2 = 260 - 6 \cdot 10^3$.

At twist, a stream by means of axial vortex generator with flat or profiles blades hydraulic resistance, ΔP_v depends only on average speed of gas in cracks between blades and their forms. Hydraulic resistance vortex generator with flat blades in 1.5 times above, than with profiles that speaks reduction of losses an input {entrance} in the device and various conditions of movement of a stream between plates (Kutepov, 1977).

At a turbulent mode of current, $Re_g \leq 5.5 \cdot 10^3$ for vortex generator with flat blades dependence is offered.

$$\Delta P_v = 18,7 \rho W_{pl}^2 Re_g^{-0,25} F_{pl} \tag{17.8}$$

$$Re_g = \frac{W_{pl} \cdot d_{ekb} \cdot \rho}{\mu};$$

where d_{ekb} – Equivalent diameter of cracks, m;

W_{pl} – Average speed of gas in cracks between plates, m/s;

F_{pl} – The relative area of alive section vortex generator, shares of a unit.

For vortex generator with spiral blades, the numerical factor in this formula is replaced on 12.7.

At a mode of the developed turbulence, $Re_g \geq 5.5 \cdot 10^3$ hydraulic resistance vortex generator is defined {determined} accordingly under formulas.

$$\Delta P_v = 1,4 \rho W_{pl}^2 \text{ (flat blades)} \tag{17.9}$$

$$\Delta P_v = 0,9 \rho W_{pl}^2 \text{ (spiral blades)}$$

17.4   CENTRIFUGAL DEDUSTERS WITH CYLINDRICAL AIR SWIRLERS OF TYPE TBVP

Centrifugal devices with vortex generator type TBVP are most full investigated {researched} in connection with use of the twirled streams in processes heat- and mass transfer, and also burning (Samsonov, 1974; and Kolborn, 1964; and Kepelman, 1960).

Investigating {Researching} separation elements tangential blade vortex generator, the author of work (Kepelman, 1960) considers {counts} as the most expedient definition ζ_v to make under the formula

$$\zeta_v = \exp\left(4,23 - 2,345\frac{F_{pl}}{F}\right),$$ (17.10)

where F_{pl} – the total area of cracks on an output {exit} from vortex generator, m^2;

F – The area of the case of a deduster, m^2.

The Estimation of influence of separate design data and a choice of the best variant vortex generator can be made on the empirical equation:

$$\zeta_v = 3,1\alpha^{0,7}\left(\frac{B}{\varepsilon}\right)^{1,54}\left(\frac{h_1}{D}\right)^{-1,48}$$ (17.11)

where α – a corner between the next plates on an input {entrance} in vortex generator, a hailstones;

B – Width of plates, m;

ε – The minimal distance between plates on an output {exit} from vortex generator, m.

The Factor of resistance ζ_v in two last formulas concerns to mean speeds in section of the device.

17.5   CENTRIFUGAL DEDUSTERS WITH CYLINDRICAL VORTEX GENERATOR TYPE TBVC

Cylindrical vortex generator with the central input of a stream within the limits of at $z = 0$ represents nozzles. It {he} has found application for catching drops and distributions of ventilating air in premises {rooms} (Anthony, 1970).

It is established {installed}, that change of relative distance between a disk and a target edge of the cylindrical channel $\hat{h} = h_1/d_1$ influences structure of an induced field of speeds and hydraulic resistance a nozzle only in a range $\hat{h} \leq 0.7$ – 0.8.

At $\hat{h} \geq 0.7$ – 0.8, the factor of the resistance, which have been counted up on speed in section of a pipe, makes $\zeta = 1$, and, hence, the disk put across the following stream, does not cause additional losses in comparison with the open pipe.

Similar dependences $\zeta_v = f(h_l)$ are received and for the twirled jet following in unlimited or limited space (Anthony, 1970). At increase, h_l up to 0.6–0.7 size ζ_v vortex generator sharply decreases, and then, at the further increase h_l practically remains to a constant.

Influence of relative diameters of an input {entrance} in vortex generator d'$_l$ = d_l/D and a disk d'$_2$ = d_2/D on hydraulic resistance drip pan is opposite and is defined {determined} by the equation,

$$\zeta = \frac{0,025}{\overline{d}_1^8} + \frac{30,5\overline{d}_2^4}{\overline{d}_1} \qquad (17.12)$$

Therefore function $\zeta_v = f(d'_1, d'_2)$ has an extremum defining {determining} the optimum attitude {relation}

$$\overline{d}_{1.opt} = 0,573 \cdot m^{-0,364} \qquad (17.13)$$

where $m = d'_2/d_l$

The Corner β_l of installation of blades on an input {entrance} in vortex generator shows on hydraulic resistance drip pan only at values $\beta_l \leq 50°$, which should be accepted as optimum (Mielazarek, 1978).

As a result of processing, the experimental data received by authors in bench and industrial conditions, for calculation of hydraulic resistance of a deduster with vortex generator type TBVC, the equation is offered,

$$\zeta_v = 3,54 \left(\frac{0,025}{\overline{d}_1^8} + \frac{30,5 \cdot \overline{d}_2^4}{\overline{d}_1} \right) \cdot \cos^{10} \beta_2, \qquad (17.14)$$

which is fair at

$$0,425 \leq \overline{d}_1 \leq 0,7; \ 0,4 \leq \overline{d}_2 \leq 0,8; \ \beta_1 \geq 50°; \ 0 \leq \beta_2 \leq 15°; \ \overline{h}_1 \geq 0,73.$$

17.6   CENTRIFUGAL DEDUSTERS WITH CONIC VORTEX GENERATOR TYPE ATVP AND ATVC

Hydraulic characteristics conic vortex generator type ATVP are most full investigated {researched} by development high-economical direct-flow cyclones (Kouzov, 1972; and Adamar, 1970).

By Researches, it is established {installed} that the increase in quantity {amount} of blades vortex generator z at F_{pl} = $const$ leads to falling ζ_v irrespective of their form. Character of dependence $\zeta_v = f(z)$ speaks smooth formation twist a stream and reduction of losses by "impact." With increase in number of blades, the degree compression separate jets considerably {much} decreases, their length and a corner of turn, and also is stabilized an axis of rotation.

The increase in the total area of cracks leads to falling ζ_v that is absolutely natural as thus entrance speed decreases. At F_{pl} = $const$, z = $const$ with reduction of relative diameter of a disk $d'_2 = d_2/D$ function $\zeta_v = f(d'_2)$ changes on hyperbolic dependence that speaks change of distribution of tangential speed on radius at aspiration W_t to a constant.

The increase in height vortex generator at constant values does not influence on ζ_v. The factor of hydraulic resistance of devices with conic vortex generator can be calculated on the equation,

$$\zeta_v = 8,5 \frac{z^{0,4} \cdot \sqrt{\overline{d'_2}}}{\overline{F}_{pl}^{/2}} \qquad (17.15)$$

which is fair at

$$\mathrm{Re} \geq 9 \cdot 10^4; \, z = 1 \div 8; \, \overline{d'_2} = 0,05 \div 0,8; \, \overline{F}_{pl} = 0,2 \div 1,0$$

Researches on hydraulic resistance vortex generator type ATVC, it was not spent.

Attempt to generalize experimental data on hydraulic resistance of various types vortex generator has been undertaken by H. O. Nurste (1973). Investigating{researching} burner devices with vortex generator type AV, TBVP, and tangential air intake, the author has established {installed}, that hydraulic resistance for all types vortex generator is defined {determined} by a parity {ratio} of the areas F/F_{out} and distance from the center of gravity of the area F_{out} up to an axis of the channel R_{out}. For blades vortex generator essential influence renders also a corner of an output {exit} of a stream β_2, so the general {common} expression of dependence of hydraulic resistance of different types vortex generator from design data can be presented in the form of

$$\zeta_v = f\left(\frac{F}{F_{out}}; \frac{R_{out}}{R}; \beta_2\right)$$ (17.16)

In the right part, it is possible to make a complex meaning the settlement dimensionless moment of quantity {amount} of movement on an output {exit} from vortex generator Of parameters:

$$\bar{M}_{out} = \frac{M_{out}}{\pi \rho W_a^2 R^3} = \frac{F}{F_{out}} \times \frac{R_{out}}{R} \times \sin \beta_2$$ (17.17)

Which on sense and the maintenance {contents} is similar to Ahmedova R. B. used in works to size n (Barahtenko, 1973) and to known parameter Rossbi. Values M_{out} at calculations turn out a little bit distinct from them.

On the basis of the theory of an ideal centrifugal atomizer and H. O. Nurste's received experimental data suggests to count hydraulic resistance vortex generator on the equation,

$$\zeta_v = \varepsilon_c \mu^{-2}$$ (17.18)

where μ – factor of the expiration vortex generator, equal

$$\mu = \left[\left(\frac{F}{F_{out}} - 1\right)^2 \sin^2 \beta_2 + \frac{2 - \psi}{\psi^3}\right]^{-0,5}$$ (17.19)

ε_c – The correction factor accepted identical for vortex generator TBVP and AV $\varepsilon_c = 1.1$.

The Factor ψ pays off from a condition of the maximal charge at the given pressure P_{out}, i.e., from a condition d μ/d $\psi = 0$:

$$\frac{1 - \psi}{\sqrt{\dfrac{\psi^3}{2}}} = \frac{F}{F_{out}} \cdot \frac{R_{out}}{R} \sin \beta_2$$ (17.20)

Hydraulic resistance of the target branch pipe ΔP_{out}, executed in the form of a flat diaphragm and a cylindrical or conic chipper, is defined {determined} basically by relative diameter compression d_3 and with a sufficient degree of accuracy can be calculated on the equation

$$\Delta P_{out} = 1,45 \rho W_{cp}^2 \cdot \overline{d}_3^{-2,94} \tag{17.21}$$

At use of a target branch pipe in the form of diffuser with the central corner of disclosing $\alpha = 8°30›$ and length $1.1·D$ hydraulic resistance of a deduster decreases for 10–12 per cent due to restoration of kinetic energy of a rotating stream.

Data Resulted {Brought} in the literature on hydraulic resistance water gas a layer ΔP_{g+p} are rather inconsistent, however, the analysis of available experimental data (Pravdin, 1973; Guhman, 1980; Kutepov, 1977; Bralove, 1981; and Ribchinskij, 1970) and the data received by authors in laboratory and industrial conditions, allow to ascertain the following.

For dedusters with counter flow tap {removal} of a film of a liquid in separation to a zone at $\acute{L} = L/G \leq 2$ hydraulic resistance does not depend on loading on a liquid and remains practically constant. At the subsequent increase, $\acute{L}$ hydraulic resistance starts to increase. Prominent feature is that fact, that at small values $\acute{L} \approx 0.5 - 1.0$ resistance of the irrigated device below dry, however with growth of relative loading hydraulic resistance comes nearer to resistance of the dry device and at $\acute{L} \geq 2$ starts to surpass it {him}.

Such dependence can be explained by significant change of structures of axial and tangential speeds of gas, and also significant clearing of turbulent pulsations of a gas stream.

At $\acute{L} \geq 2$ value of factor of hydraulic resistance of dedusters are equal:
For dedusters with axial vortex generator with flat blades

$$\zeta_{0V} = 2,44 \left(\frac{L}{G} \right)^{0,2};$$

For dedusters with axial vortex generator with spiral blades

$$\zeta_{0V} = 1,56 \left(\frac{L}{G} \right)^{0,2};$$

For dedusters with cylindrical vortex generator type TBVP

$$\zeta_{0V} = 5,2 \left(\frac{L}{G} \right)^{0,4}.$$

The Factor of resistance ζ_v concerns to speed of air in free section vortex generator.

At tap {removal} of a film of a liquid in the top part of the case of a deduster (turned parallel flow phases) decrease {reduction} in hydraulic resistance of the irrigated device occurs {happens} until the condition $\acute{L} \leq 0.5 - 0.6$ is observed.

At $\acute{L} \geq 0.6$, the size ζ_v increases proportionally to growth of this attitude {relation} due to increase in expenses at transportation of a liquid.

Factors of hydraulic resistance for such type of dedusters can be calculated on the equations:

For dedusters with two entry axial vortex generator

$$\text{Re}_g > 17800, \frac{S}{d_2} = 1$$

$$\zeta_{0V} = 8,8\,\text{Re}_l^{0,16} \text{ at } \text{Re}_l < 1200\,;$$

$$\zeta_{0V} = 0,8\,\text{Re}_l^{0,5} \text{ at } \text{Re}_l < 1260\,;$$

For dedusters with cylindrical vortex generator type TVCP at $\acute{L} \leq 0.6$

$$\zeta_{0V} = 5,2\left(\frac{F_{\acute{o}}}{F}\right)^{-1,4} \cdot \left(10\frac{L}{G+L}\right)^{-0,08} \tag{17.22}$$

at $\acute{L} \geq 0.6$

$$\zeta_{0V} = 0,904\frac{L}{G} + 8,2$$

The resulted {brought} equations are fair at $F_{pl}/F = 0.4 - 1.2$ and $Re_g \geq 2 \cdot 10^4$; factors of resistance are carried to mean speeds in separation to a zone.

In case of descending vortical movement of phases, the body weight of a liquid promotes rectification twist and to reduction of size of absolute speed at a constant axial. Rectification twist affects essentially at small speeds of gas, significant loadings on a liquid and weak twist. The difference in hydraulic resistance of dedusters with an ascending and descending stream decreases with growth of speed of gas and at values $Re_g \geq 30,000$ aspires to zero (Ribchinskij, 1970).

$$\mathrm{Re}_g > 17800\,\frac{S}{D} = 1$$

For dedusters with two entry axial vortex generator at factors of hydraulic resistance are defined {determined} by the equations (Mielazarek, 1978):

$$\zeta_{0vor} = 7,2\,\mathrm{Re}_l^{0,16} \text{ at } \mathrm{Re}_l < 1200\,;$$

$$\zeta_{0vor} = 0,676\,\mathrm{Re}_l^{0,5} \text{ at } \mathrm{Re}_l > 1200\,;$$

For dedusters with conic vortex generator (Adamar, 1970) equation is offered

$$\Delta P = 9,8 \cdot \rho_g \cdot W_{cp}^2 \left(16,1\cdot 10^{-6}\,\overline{L}\cdot \mathrm{Re}_g^{-1}\,\overline{H} + 1,21\right) \tag{17.23}$$

where $\acute{L}$ – specific loading on a liquid, l/m³;
$\hat{H}$ – Relative length of the moistened surface of the case.

17.7   CONCLUSIONS

The analysis of known builds of rotary connections has allowed to reveal major factors (loading on a liquid phase, a monkey wall and a stream breakaway, discharge connection sizes), working upon a water resistance of whirlwind apparatuses. Resistance is defined by the sum of pressure losses of an air swirler, in the separation chamber and on an exit from it. It is shown that pressure losses in an air swirler depend on its geometry and extent of a twisting of gas.

Dependences for calculation on a water resistance in various types of air swirlers are offered: vaned, tangential, scroll, with spiral and flat guide vanes.

KEYWORDS

- **Flat guide vanes**
- **Spiral guide vanes**
- **The conic air swirler**
- **The cylindrical air swirler**
- **The water resistance**
- **Water resistance factor**

REFERENCES

Abramovich, G. N.; Applied fluid dynamics. M.: Publishing House the Tehniko-Theoretical Literature; **1953.**

Pravdin, V. A.; Aerohydrodynamic streams. *Chem. Eng. Ind.* **1973,** *10,* 35–37, (in Russian).

Mustashkin, F. A.; Two-phase streams. Kazan: Works КCTU; **1970,** *45,* 94–99, (in Russian).

Karpenkov, A. F.; The dust separation theory in the turbulent wet washer. Works КCTU; **1970,** *45,* 40–45, (in Russian).

Nurste, H. O.; Ref. Diss. Cand. Tech. Sci. Institute Warmly—to Electrophysicists an Academy of Sciences of Est. Tallinn: The Soviet Socialist Republic; **1973.**

Fucs, N. A.; Mechanics of Fluids. **1955.**

Nikolaev, N. A.; and Larks, H. M.; Theory Basis Chemical Technology. **1973,** 3.

Ovchinnikov, A. A.; Research of hydroaerodynamic laws in vortical mass transfer the device with tangential blade vortex generator. In: Auto Abstract. Kazan: Institute of Chemical Technology; **1973.**

Korotkov, J. F.; and Nikolaev, N. A.; High schools. *Chem.* **1972,** 5.

Hollow, V. F.; Collection: "Questions of Designing and Mounting Sanitary Systems." **1972,** *32,* 22–26.

Schukin, V. K.; Heat exchange and hydrodynamics of internal streams in fields of mass forces. M.: "Engineering Industry." **1970.**

Pustotaja, V. F.; In the Collector: "Questions of designing and installation of sanitary systems." **1972,** *32,* 22–26 (in Russian).

Barahtenko, G. M.; In the Collector: "Cars and apparatuses of food manufactures." Sverdlovsk; **1971,** *1,* 53–62 (in Russian).

Ershov, E. I.; et al. New builds of separators for clearing of industrial gases. The Survey Information. **1973.**

Margolin, E. V.; and Prihodko, V. I.; Perfection of production engineering of wet clearing of gas emissions at aluminium factories. M.: "Engineering Industry." **1977.**

Guhman, L. M.; and Minakov, V. V.; Cars and the oil equipment. M.: Higher School. **1986,** *9,* 20–22 (in Russian).

Bunkin, V. A.; Working out and research of the mass-transfer apparatus with direct-flow rotary connections. Ref. Diss. Cand. Tech. Sci. Kazan: КCTU; **1970.**

Samsonov, V. T.; In the Collector: "Ventilation and an air conditioning at the polygraphic factories." Works Scientific Research Institute a Polygraph, M.: "Book." **1974.**

Kolborn, J.; and Land, C. U.; Heat Mass Transfer. **1964,** *7,* 1170.

Kepelman, M. N.; and Eskin, N. B.; Design procedure of horizontal film and centrifugal separators. Data Centre. **1960.**

Kutepov, A. M.; et al. Hydrodynamics and heat exchange at steam formation. M.: "Higher School." **1977.**

Lang, C. U.; Heat mass transfer. *Chem. Technol.* **1964,** *7,* 1195.

Marchall, J.; Heat mass transfer. *Chem.* **1977,** *20,* 227.

Bralove, A. L.; Dust separation equipment. Part II, *5,* **1981.**

Anthony, A. W.; United States Technical Conference on Air Pollution. **1970,** 310.

26. Kouzov, P. A.; and Milnikov, S. I.; Works of Institutes of a Lab Our Safety of the All-Union Central Council of Trade Unions. **1972,** *77,* 6–12 p.

Adamar, V. A.; *Chem. Technol.* **1970,** *12,* 1141–1143 p.

Ribchinskij, S. J.; The survey information. "Bowels." **1970,** *4,* 8.

Johnstone, H. F.; Field, R. B.; Brink, W.; and Tassler, M. S.; *Ind. Eng. Chem.* **1964,** *46(8),* 1604.

Mielazarek, M.; and Koch, R.; *Chem. Eng. Jpn.* **1978,** *11(6),* 469.

Newman, A. V.; *Trans. Amer. Inst. Chem. Eng.* **1946,** *42,* 79.

Ahmedov, R. V.; Aerodynamics of the twirled jet. "Energy." **1977.**

Kutepov, A. M.; and Nepomnyaschy, E. A.; Theory Basis Chemical Technology. **1973,** 6.

Kutepov, A. M.; and Nepomnyaschy, E. A.; Research of hydrodynamics. *Chem. Chem. Technol.* **1976,** *1,* 21.

Kutepov, A. M.; Hydrodynamics and heat exchange at steam formation. M.: "Higher School." **1977**

Jofinov, G. A.; Collection: Institutes of a Lab Our Safety of the All-Union Central Council of Trade Unions. **1975,** 96.

Jofinov, G. A.; Collection: Institutes of a Lab Our Safety of the All-Union Central Council of Trade Unions. **1975,** 97.

Uzhov, V. N.; and Valdberg, A. J.; Clearing of gases wet filters. "Chemistry." **1973,** 247.

Zwechek, E.; Report on a symposium "Ways of clearing of prom. Gases and the equipment for these purposes." Zaporozhye. **1976.**

CHAPTER 18

EFFECT OF CONCRETION ON PROCESS OF SEDIMENTATION OF CORPUSCLES OF A DUST

CONTENTS

18.1 INTRODUCTION

Aerosols, as well as many other disperse systems, have the restricted life time. In them, there are the various processes leading to integration of primary corpuscles, to their aggregation, formation of offsprings, and the subsequent sedimentation. Some of these processes proceed spontaneously, others—under the influence of electric, hydrodynamic, or a gravitational field. In the course of integration of corpuscles, there is a decrease in superficial energy at the expense of decrease of their specific surface [1–12].

Integration of corpuscles can go two ways: as a result of isothermal distillation, if corpuscles liquid (Calvin's effect), and as a result collisions and an adhesion-Concretions. In case of merge of liquid corpuscles, this process name *coalescence*. So, generally *as concretion* is called decrease of degree of dispersion, i.e., decrease of number of corpuscles of a dispersoid at their integration [13–18].

The Brownian motion of corpuscles can be the concretion reason and in this case it is called *as Brownian, either a spontaneous coagulation,* or affecting of superposed forces—hydrodynamic, electric, gravitational, etc., corpuscles leading to collision, and in this case concretion is called *as forced.* The basic regularity of concretion of aerosols was installed for the first time by the Polish physicist M. Smoluhovsky. For the account of efficiency of an adhesion of corpuscles at collision, it has injected two concepts: "racing" and "slow" concretion. In the first case of collision conduct to an adhesion of corpuscles, and in the another adhesion occurs not from the first collision or not all faced corpuscles coagulate[19–28].

In the first case of collision conduct to an adhesion of corpuscles, and in the another adhesion occurs not after the first collision or not all faced corpuscles coagulate[29–31].

18.2 COAGULATION OF EQUIGRANULAR SPHERICAL CORPUSCLES

Most simple aspect of concretion is a thermal coagulation of equigranular spherical corpuscles. As are observed, only the first some certificates of collision of corpuscles, the size of the formed corpuscles will not differ aloud from a size of initial corpuscles[32–40]. This model is used many years for firm corpuscles and can form a basis for definition of a constant of concretion. It approaches and for the description of concretion of drops of a liquid as the size of drops after merge increases proportionally to a cube root from quantity of drops, its components [41]. This approach was offered Smoluhovsky(1936a; 1936b) for concretion in the diluted electrolytes, but it can be used for aerosols within the restrictions observed above (Saltanov, 1936; and Hidy, 1965). In approach Smoluhovsky, it is supposed that in system of spherical corpuscles in diameter $2R$ distances between corpuscles are chaotically distributed, at $\tau = 0$. If corpuscles move also chaotically by thermal diffusion it is necessary to know probability of their collision during some time(Figure 18.1).

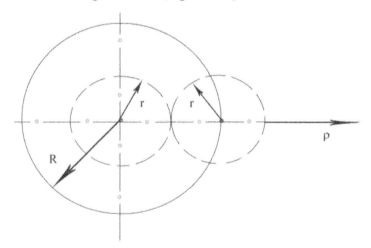

FIGURE 18.1   The concretion loading diagram.

Smoluhovsky, the first has observed a case when one corpuscle fixed in space, is the center of concretion for other corpuscles. It has defined speed of diffusion of other corpuscles to this central corpuscle. The equation of an unsteady-state diffusion looks like:

$$\frac{\partial C}{\partial \tau} = D_d \nabla^2 C \tag{18.1}$$

whereC—concentration of corpuscles and D_d—factor of their diffusion. If r—the distance from the center of the fixed corpuscle assuming spherical symmetry of system, Eq. (18.1) can be written down the equation in an aspect:

$$\frac{\partial C}{\partial \tau} = D_d \left(\frac{\partial^2 C}{\partial r^2} + \frac{2}{r} \frac{\partial C}{\partial r} \right)$$

(18.2)

or in more convenient form:

$$\frac{\partial (Cr)}{\partial \tau} = D_d \frac{\partial^2 (Cr)}{\partial r^2}$$

(18.3)

As corpuscles have the same size, we assume that they will face the central corpuscle when pass in distance limits $2R$ from it (Figure 18.2). Thus, concentration will be equal in this point to null, that is $C' = 0$ at $r = 2R$ (for $\tau = 0$). Besides, it was originally supposed that corpuscles are in regular intervals distributed on all volume with concentration C. Thus, $C' = C$ at $\tau = 0$.

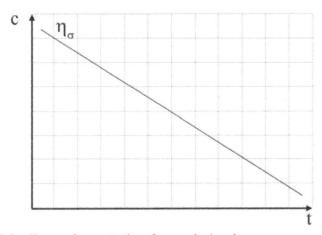

FIGURE 18.2   Change of concentration of corpuscles in a time.

Then Eq. (18.3) it is possible to present the equation:

$$C' = C \left[1 - \frac{2R}{r} + \frac{2R}{r} erf \left(\frac{r - 2R}{2\sqrt{D_d \tau}} \right) \right]$$

(18.4)

The number of corpuscles N, which diffuses within distance $2R$ from the central fixed corpuscle in unit of time, is equal to product of a diffused stream on the square of a surface of sphere in radius $2R$. The diffused stream is defined from the equation

$$J = -D_d (\partial C'/\partial r), \tag{18.5}$$

Where the derivative (dC'/dr) should be sized up at $r = 2R$. Thus,

$$\vec{N} = 16\pi R^2 D_d \frac{\partial C'}{\partial r}. \tag{18.6}$$

At $r = 2R$. And from the Eq. (18.4) we will gain

$$\frac{\partial C'}{\partial r} = \frac{C}{2R}\left[1 + \frac{2R}{\sqrt{\pi D_d \tau}}\right] \tag{18.7}$$

Then the number of corpuscles, which attain a surface surrounding the central corpuscle, in time $d\tau$, makes

$$\vec{N}d\tau = 8\pi R D_d C\left[1 + \frac{2R}{\sqrt{\pi D_d \tau}}\right]d\tau. \tag{18.8}$$

Now, we will assume that the fixed corpuscle can diffuse the same as other corpuscles. Diffusion also should be considered[42–44]. The general factor of diffusion of two corpuscles rather is each other equal to the sum of factors of diffusion of separate corpuscles so the propelled corpuscle faces with

$$16\pi R D C\left[1 + \frac{2R}{\sqrt{\pi D \tau}}\right]d\tau$$

Corpuscles in the interval $d\tau$, in view of that the Eq. (18.8) is applicable for equal sizes. In unit volume will occur $C/2$ collisions, if C—number of corpuscles two corpuscles participate in unit volume and in each collision. The number of collisions in unit volume, which occur in time $d\tau$, is presented by the equation:

$$\frac{dC}{d\tau} = -\frac{16}{2}\pi R D_d C^2\left[1 + \frac{2R}{\sqrt{\pi D_d \tau}}\right]. \tag{18.9}$$

The another member in brackets, it is possible not to consider, as it will be much less unit if τ it is great enough. The Fuchs (1955; and 1951) has shown that

$$2R/\sqrt{\pi D_d} = \xi$$

The probability of that a corpuscle originally is near to the fixed corpuscle. Thus, ζ it is aimed to null as stationary speed increases. As, ζ it is not enough for practical conditions, this magnitude neglect. However, it is necessary to mean that in some conditions ζ, it is necessary to consider[45–49]. This member leads to increase in speed of concretion. Defining a constant of concretion K_0 as

$$K_0 = 16\pi R D_d = \frac{8kT}{3\mu_c}\zeta_c \qquad (18.10)$$

We gain results which, at least, for large corpuscles will not depend on a size of corpuscles.

Using K_0 as a constant of concretion and neglecting the another member in the Eq. (18.9), we will gain the usual form of the equation of concretion:

$$\frac{dC}{d\tau} = -\frac{K_0}{2}C^2 \qquad (18.11)$$

Integrating the Eq. (18.11) at entry condition $C = C_0$ when $\tau = 0$, we will gain:

$$\frac{1}{C} - \frac{1}{C_0} = \frac{K_0}{2}\tau. \qquad (18.12)$$

The Eq. (18.12) shows that the reciprocal quantity of concentration of corpuscles is a time linear function, and the straight line inclination gives a concretion constant. If τ' assignas a time for which concentration decreases twice,

$$C = C_0/\left(1 + \tau/\tau^*\right). \qquad (18.13)$$

Expression (18.13) can be directly used for machining of experimental data on concretion.

18.3   DUST LAYING ON DROPS AT LIQUID SPRAYING

Efficiency of sedimentation of corpuscles on drops of liquid (the kinematic concretion) depends first of all on magnitude of their relative traverse speed w. The kinematic (gravitational) concretion to proceed at free falling of drops through motionless an aerosol of count concentration of n[50–54].In this case, number of the small corpuscles entrained by one drop in 1s, it can is possible to define by formula

$$Q = 1/4\, n\pi d_K^2 v_c \eta_\Sigma \qquad (18.14)$$

If drops are precipitated in a moving stream of the aerosol, which speed it is impossible to neglect Eq. (18.14), it is necessary to inject a relative traverse speed into the formula w corpuscles concerning a drop instead of speed of subsidence v_e*

The total factor of capture of corpuscles a spherical drop η depends from. A flow regime.

Efficiency of trapping according to Fuchs calculations is defined first of all by a size of corpuscles. For example, for corpuscles in density $p = 2{,}000$ kg/m³ those are trapped only, which-g size > 0.5 μm. Thus for corpuscles of small sizes (0.5–0.7 μm), the more largely a drop, the efficiency of trapping above. It proves to be true only in a case small relative speeds.

If relative speeds are great, as it occurs at liquid injection in a gas stream efficiency of concretion of corpuscles grows with decrease of a size of drops. It is possible to be convinced of it, if expression (18.14) to refer to drop volume.

Capture of corpuscles by drops depends on the several reasons. Here, along with the kinematic proceeds and graded region concretion. The momentous role in sedimentation is played by turbulence of a stream.

In wet-type collectors of bubbling type motion of gas and liquid drops can be organized on one of three circuit designs: counterflow, direct-flow, or with a cross current (Valdberg, 1969; and 1971). We will carry out the analysis of process of sedimentation of a dust on drops of liquid depending on the circuit design of motion of streams and it is definable its efficiency.

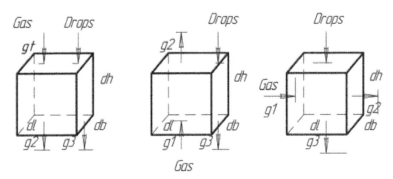

FIGURE 18.3 The loading diagram of efficiency of the apparatus for motion: (a)—a forward flow; (b)—a backward flow; and (c)—a cross current.

Let's gate out an element of space with a size *dldbdh*(Figure 18.3). We will mark out a heading of material streams depending on the motion circuit design. In all three cases on the overhead basil *(dldb), the* stream of the drops, which volume flow rate arrives, m³/s,

$$V_l = \alpha v \left(dldb \right)$$

(18.15)

Where *v*—a traverse speed of drops; and—a share of the volume occupied with drops.

The dust-laden flow of gases depending on the motion circuit design arrives on overhead, bottom, or a lateral face *(dbdh)*.

If mass concentration of a dust on an entry in a volume element is equal in gas C, kg/m³ the dust stream on an entry will make, kg/s:

For direct-flow and counterflow circuit designs

$$g_1 = u \left(dldb \right) C$$

(18.16)

For the circuit design about a gas cross-flow

$$g_1 = u \left(dbdh \right) C$$

(18.17)

where *u*—speed of a dust, equal to speed of gas.

On an exit from a volume element gas will contain a dust *(C~dC)*, the kg/m³, and a dust stream will make accordingly, kg/s:

$$g_2 = u\left(dldb\right)\left(C - dC\right) \qquad (18.18)$$

For the circuit design about a gas cross-flow

$$g_2 = u\left(dbdh\right)\left(C - dC\right) \qquad (18.19)$$

Dust laying on drops occurs to relative speed w. In the circuit design about a forward flow this speed $w=u - v_t$ backward flow $w=u - v_9$ and in case of cross motion in a gas stream heading

$$\omega_l = u, v_l = 0$$

According to the formula (18.15) on one drop in diameter d_K at efficiency of capture t_s, it is precipitated corpuscles in weight:

$$1/4\, Cw\pi d_K^2\, \eta_\Sigma \qquad (18.20)$$

In total in the gated out volume, there is a thaw

$$6\left(dldbdh\right)\alpha/\left(\pi d_K^3\right) \qquad (18.21)$$

Thus, as a result of capture by drops a dust particle flux

$$g_3 = \frac{6\pi d_K^2\, \alpha\omega C\, \eta_\Sigma}{4\pi d_K^3}\left(dldbdh\right) = \frac{3}{2}\cdot\frac{\alpha\omega C\, \eta_\Sigma}{d_K}\left(dldbdh\right) \qquad (18.22)$$

Let's write down the equation of a material balance of a volume element:

$$g_1 - g_2 - g_3 = 0$$

for direct- and the counterflow circuit design

$$uCdldb - u\left(C - dC\right)dldb - \frac{3}{2}\cdot\frac{\alpha\omega C\, \eta_\Sigma}{d_K}dldbdh = 0 \qquad (18.23)$$

and for a cross current

$$uCdbdh - u(C - dC)dbdh - \frac{3}{2} \cdot \frac{\alpha\omega C\eta_\Sigma}{d_K} dldbdh = 0 \qquad (18.24)$$

After transformations, we will gain for the first two circuit designs, where

$$\omega = u \pm v$$

$$\frac{dC}{C} = -\frac{3}{2} \cdot \frac{\alpha\eta_\Sigma}{d_K} \cdot \frac{\omega}{u} dh \qquad (18.25)$$

And for a cross current, where $w=u$ at $u=0$,

$$\frac{dC}{C} = -\frac{3}{2} \cdot \frac{\alpha\eta_\Sigma}{d_K} \cdot dl \qquad (18.26)$$

Let us inject a parameter of an irrigation of gases m, m³/m³, as the relation of volume flow rates of a liquid and gases: for the first two circuit designs

$$m = \frac{V_l}{V_g} = \frac{\alpha v dldb}{u dldb} = \frac{\alpha v}{u} \qquad (18.27)$$

or

$$\frac{\alpha\omega}{u} = m\frac{\omega}{v}$$

And for the circuit design about a cross-flow in terms of that $u = v$

$$m = \frac{V_l}{V_g} = \frac{\alpha dl}{ah} \qquad (18.28)$$

or $\alpha dl = mdh$

Let us substitute the gained expressions in the Eqs. (18.25)–(18.26):

$$\frac{dC}{C} = -\frac{3}{2}m\frac{\omega}{v}\frac{\eta_\Sigma}{d_K}dh$$

$$\frac{dC}{C} = -\frac{3}{2}m\frac{\eta_\Sigma}{d_K}dh \tag{18.29}$$

After integration on all altitude of a zone of contact of a dust with drops of liquid (apparatus altitude) expression (18.29) can be written down in an aspect.

For circuit designs with line and counterflow motion and

$$\eta = 1 - \exp\left(-\frac{3}{2}m\frac{\omega}{v}\cdot\frac{H}{d_K}\eta_\Sigma\right) \tag{18.30}$$

For the circuit design with a lateral motion of streams.

$$\eta = 1 - \exp\left(-\frac{3}{2}m\cdot\frac{H}{d_K}\eta_\Sigma\right) \tag{18.31}$$

The analysis of efficiency of three circuit designs has been spent at following assumptions:

(1) drops are distributed in regular intervals by volume;
(2) process isothermal, without changes of phase;
(3) corpuscles have speed of gas and are trapped only by drops as a result of collisions in the presence of relative speed.

Real process of trapping of a dust by drops of liquid much more difficult also does not give in to the full account. Therefore, efficiency of the apparatus defines as a rule, by means of empirical dependences.

For concretion of corpuscles of two various sizes: it is possible to use the same approximation, as in case of a monodisperse aerosol. Except for replacement $2R$ on $R_1 + R_2$ and $2D_d$ on $D_{d1} + D_{d2}$. Then K_{12}, it will be equal

$$K_{12} = 4\pi\left(R_1 + R_2\right)\left(D_{d1} + D_{d2}\right). \tag{18.32}$$

Dependence of a constant of concretion through mobility of corpuscles looks like:

$$K_{12} = 4\pi\left(R_1 + R_2\right)\left(B_1 + B_2\right)kT. \tag{18.33}$$

$B = D_d/(kT) = \zeta_c/(6\pi\mu R)$—mobility of a corpuscle.

The concretion constant has the minimum value in case of corpuscles of equal sizes.

18.4   COAGULATION OF UNEQUIGRANULAR SPHERICAL CORPUSCLES

18.4.1   COAGULATION OF UNEQUIGRANULAR CORPUSCLES

Tikhomirov, Tunitsky, and Petrjanov(Watson, 1936), also have shown that if it is necessary to know change only concentration of corpuscles$dC/d\ \tau$ concretion constant was possibly to merge in one K, ' which will be expressed through average values

$$\langle R \rangle, \ \langle 1/R \rangle \ \text{and} \ \langle 1/R^2 \rangle$$

$$K^* = \frac{4kT}{3\mu_c}\left[1 + \langle R\rangle\langle 1/R\rangle + Al_d\,\langle 1/R\rangle + Al_d\,\langle R\rangle\langle 1/R^2\rangle\right]. \tag{18.34}$$

Here A—the correction of Kenningem, equal 1.35 and l_d—a free average way of a corpuscle to gas. Thus, K' it is possible to use as K_0 in Eq. (18.11).

If it is primary sizes of corpuscles obey the law logarifmicheski-normal distribution expressions for K' can be written down in compound diameter $2R_g$ and a standard geometrical deviation σ_g.

$$K^* = \frac{4kT}{3\mu_c}\left[1 + \exp\{\ln^2\sigma_g\} + \frac{Al_d}{R_g}\exp\left\{\frac{1}{2}\ln^2\sigma_g\right\} + \frac{Al_d}{R_g}\exp\left\{\frac{5}{2}\ln^2\sigma_g\right\}\right] \tag{18.35}$$

With increment of polydispersity, the constant of concretion K' can become enough big, considering that limiting values of sizes of corpuscles sweepingly increase owing to the joint of corpuscles, especially smaller sizes. It is obvious that speed of concretion for unequigranular an aerosol more than for equigranular. However, monodisperse aerosol concretion leads in the beginning to polydispersity increase so for any coagulating an aerosol the concretion constant is not constant, and depends on speed of concretion. It is no wonder that interpretation of the data on concretion is difficult, but knocks that fact that simple theory of concretion stated here is suitable for the description of many cases.

18.4.2   REFINEMENT OF A CONSTANT OF CONCRETION

Experimental value of a constant of concretion can be appreciable above the theoretical. On all visibilities, it is connected by that the resulted reasonings are lawful in the event that aerosol corpuscles are in regular intervals distributed by volume. Thus, uniformity of distribution depends on total number of corpuscles in system N_0 of their size V_0, a size of control volume V_k, and total amount systems V, which can be accepted equal to unit.

Let N the greatest possible number of aerosol corpuscles, which can be placed in system, and N_c average of corpuscles, which can be in control volume. Between N_c and average concentration of aerosol corpuscles C, there is a following communication:

$$N_C = V_k \langle C \rangle = \alpha V \langle C \rangle. \tag{18.36}$$

Local value of number of corpuscles in control volume is connected with their local concentration as follows:

$$n = V_k C = \alpha V C \tag{18.37}$$

By definition $C = N_0/V$. On the other hand, $V = NV$ hence:

$$\frac{N_0}{N} = \langle C \rangle V_0. \tag{18.38}$$

Then with reference to a concretion problem, it is possible to write down following expression for probability of distribution of number of corpuscles on separate elements of the system, which size is equal to control volume:

$$p(C) = \frac{1}{\langle C \rangle \sigma \sqrt{2\pi}} \exp\left\{ -\left[\frac{C - \langle C \rangle}{\langle C \rangle \sqrt{2\sigma}} \right]^2 \right\}, \qquad (18.39)$$

where σ—a dispersion

$$\sigma = \frac{1}{\sqrt{2\pi e}} \left(V_0 \langle C \rangle \right)^{n_d} \mathrm{h} \left(V_0 \langle C \rangle \right)^{n_d} \qquad (18.40)$$

and

$$n_d = \frac{3}{\ln(V/V_0)} - \frac{3}{\ln(Vk/V_0)}. \qquad (18.41)$$

Our problem consists in sizing up, as irregularity of distribution of aerosol corpuscles by volume disperse system can affect total speed of concretion. For this purpose, it is necessary to count this speed in each control volume in terms of possibly in it concentration of corpuscles.

In total such control volumes $N_k = V/V_k$. Then in $N_k\, p\,(C)$ control volume concentration of aerosol corpuscles is equal C, and speed of concretion matching to it will be defined as follows:

$$\eta = \frac{dC}{d\tau} = -\frac{K_0}{2}(C)^2 = \eta_0 \left(\frac{C}{\langle C \rangle} \right)^2,$$

Where η_0—the speed of concretion matching to average concentration C. Then, it is possible to count some factor equal to the relation of average speed of concretion to speed of concretion at average concentration:

$$\overline{\eta} = \frac{\eta}{\eta_0} = \int_0^\infty \left(\frac{C}{\langle C \rangle} \right)^2 p(C)\, dC. \qquad (18.42)$$

Actually ή is the correction of a constant of speed of concretion to the equation

$$\frac{d\langle C \rangle}{d\tau} = -\frac{K_0}{2}\overline{\eta}\langle C \rangle^2. \qquad (18.43)$$

To take advantage of the offered model, it is necessary to set magnitude of control volume or some linear scale. We will accept in the capacity of such scale a pseudo-average free way of an aerosol corpuscle

$$l_B = \tau_0 w = \tau_0 \sqrt{8kT / (\pi m_d)}, \tag{18.44}$$

Which for a corpuscle in diameter 0.01 μm will make 0.027 μm. As have shown calculations l_b much less, than it is necessary for reception of physically correct results. At conducting of calculations magnitude of a scale line has been accepted equal $L = 0.000354$ m that matches to control volume

$$V_k \approx 2 \cdot 10^{-9} m^3 .$$

Concretion of aerosol corpuscles of water in air was studied at temperature 20°C. Calculations were spent at following values of physical parameters, $\rho_d = 1{,}000$ kg/m³, $\rho_c = 1{,}200$ kg/m³, $\mu_c = 1.2 \cdot 10^{-5}$ N/m²for corpuscles of sizes $R = 10^{-10} \div 10^{-6}$ m. At small sizes of aerosol corpuscles, the correction differs from unit very little. For corpuscles of the big size calculations were not spent, as for them usability of the observed mechanism of concretion demands a special substantiation.

18.5 CALCULATION OF COAGULATION IN MOVING MEDIUM

18.5.1 FEATURES OF CONCRETION IN MOVING MEDIUM

The Brownian motion was above observed only. Coagulation can occur also in fast-moving air streams, and it is necessary to expect that at motion of corpuscles concretion will increase. There are two types of such motion. In first, there can be an arranged in sequence stream, in which corpuscles a move in one heading with various speeds, as, for example, at sedimentation unequigranular an aerosol by gravity under invariable conditions of a current. Secondly, motion can be disorder, as, for example, in case of turbulent mixing (Levitch, 1959a; Oseen, 1927; and Watson, 1963).

At first, we will observe a concretion case at the arranged in sequence motion when the large corpuscle moves through a cloud of much smaller corpuscles. How concentration of small corpuscles will decrease at affecting of the large? Admitting that speed of the arranged in sequence motion of small corpuscles is much less than speed of a large corpuscle, it is possible to consider that the large corpuscle in unit of time entrains all small corpuscles concluded in volume.

$$\pi R_2^2 w_0 \left(R_2 \right)$$

Member R_2—a size of a large corpuscle and w_0 (R_2)—the installed speed of falling of a single corpuscle. However, some small corpuscles can move with a stream aside from a front of a large corpuscle at its flow. Thereof only some share of corpuscles being in volume will be trapped by a large corpuscle. By consideration of a current round, a large corpuscle it is supposed that it is in the form of spheres. Knowing this share ε, named efficiency or in capture factor, it is possible to find quantity of the small corpuscles N trapped by a large corpuscle in unit of time

$$N = \pi \varepsilon R_L^2 w_L C,$$

where C—number of small corpuscles in unit volume.

For corpuscles about an equal size forecasting of a field of a current is extremely complicated, because the combined streams of both corpuscles should be observed. As it was specified above, current this field will change in process of rapprochement of corpuscles and also if a Reynolds number and relative sizes of corpuscles varies. Attempt of modeling of this situation for a viscous stream has been made by Hoking(1959) who has come to conclusion that between corpuscles approximately, equal sizes it is not necessary to consider collisions for corpuscles in diameter less than 36 μm. Given Hoking have been subjected criticism, and now consider that small mutual effect exists.

Coagulation of corpuscles can meet in such arranged in sequence stream what originates in an acoustic field. Here, concretion of corpuscles occurs owing to difference in traverse speeds of corpuscles of different sizes, attraction air forces between corpuscles and a radial thrust, which moves corpuscles to antinodes of waves. A complete theory of acoustical concretion till now it is not created.

For the concretion called by turbulence of a stream, it is necessary to observe two cases. First, if a pulse of corpuscles of an aerosol approximately same as at medium, they will move approximately with the same speed as well as the sections of air surrounding them. In this case, motion of corpuscles can be presented by means of factor of turbulent diffusion D. This factor can matter, in 10^4–10^6 times more than factors of thermal diffusion. Processes of the concretion called by turbulence of a stream, can be observed as usual concretion, but with use of the big factors of diffusion.

The another case of concretion of an aerosol in a turbulent stream is characterized by origination of the inertia differences between corpuscles of different sizes. Owing to turbulence of a corpuscle are sped up till the various speeds depending on a size, and can then face with each other. For a monodisperse aerosol this mechanism has no value. For unequigranular an aerosol with known size distribution of Levich(1959) has shown that speed of concretion is proportional to a basic speed of a turbulent stream in extent 8/3 that is speed of concretion increases very sweepingly with increase in speed of a turbulent stream. As very small corpuscles are sweepingly sped up, value of this mechanism decreases with decrease of a size of corpuscles, and it is the most momentous for corpuscles, whose diameters make $10^{-7} \div 10^{-6}$ m. In all cases diffusion of Brouna when diameters of corpuscles 10^{-8} m there are less predominates, that confirm the calculations carried out by us.

18.5.2 CAPTURE RATIO

Let two spherical corpuscles (drop) of commensurable sizes a move in not indignant medium under the influence of a gravity. Radiuses of these corpuscles, we will mark out through R_1 and R_2, and we will assume that $R_1 \leq R_2$. The radius of a small corpuscle is considered so big that it was possible to neglect its Brownian wandering. If mutual effect of hydrodynamic fields of corpuscles did not affect their motion in medium the small drops would face the big drop, being more low it in the cylinder, which radius is equal $R_1 + R_2$. Expression for a concretion constant thus would have the following appearance:

$$K_{1,2} = \pi \left(R_1 + R_2\right)^2 \left[w_0\left(R_2\right) - w_0\left(R_1\right)\right] \tag{18.45}$$

However, mutual effect of hydrodynamic fields of corpuscles and possibly interacting between them lead to a bending of their paths in the field of a gravity. Therefore generally

$$K_{1,2} = \pi (R_1 + R_2)^2 \left[w_0(R_2) - w_0(R_1) \right] \varepsilon(R_1, R_2),$$ (18.46)

The deviation of a capture cross-section from geometrical, called by a mutual bending of paths of corpuscles is characterized by magnitude $\varepsilon(R_1, R_2)$, which usually is called as capture factor.

Not each collision can lead to merge of drops and more logical magnitude ε to name factor of collisions (collisions) as it is accepted, for example, in Ref. (Saltanov, 1972). However, in connection with a wide circulation in the domestic and foreign literature of the term capture ratio, its renaming is not expedient. Everywhere at the theoretical analysis, it is a question only of collisions, instead of about an adhesion (concretion) of corpuscles that, it is final not same. Researches of probability of an adhesion of corpuscles after their collision, it is spent while a little, full clearness in it a question is not present. It is possible to note only that, apparently, in the majority of the situations realized at evolution of disperse systems, the probability of an adhesion of corpuscles is close to unit.

Let us assume that sizes of drops are essentially various $R_1 \ll R_2$. Then, it is possible to consider that the small corpuscle moves simply in a hydrodynamic field big, and at definition of force of resistance of medium to its motion by inhomogeneity of this field it is possible with sufficient accuracy to neglect. If the distance between surfaces of drops several times is more R_1, it is possible to neglect also forces of hydrodynamic interacting of moving sphere with a motionless flat wall. These assumptions allow to present the equation of motion of a small corpuscle in an aspect:

$$St_c \frac{d\vec{w}}{d\tau} = (\vec{v} - \vec{w}) \left[1 + \varsigma_c \left(\mathrm{Re}^* |\vec{v} - \vec{w}| \right) \right],$$

$$\vec{w} \to \vec{e} \; npu \; \tau \to -\infty,$$

$$St_c = \frac{2}{9} \frac{\rho_d R_1^2}{\mu_c R_2} w_0(R_2), \quad \mathrm{Re}^* = \frac{R_1 \rho_c w_0(R_2)}{\mu_c}.$$ (18.47)

Here:
ĕ is the individual vector directed in parallel to an axis of abscissa (on a vertical upward);

Stokes St_c is number (parameter of the inertia collision);

$w_0(R_2)$ is a vector of a hydrodynamic field of the big corpuscle in a coordinate system rigidly connected with it.

Passing round formally resulted equation of motion of a small corpuscle up to physical contact of both corpuscles, we come to a problem of purely inertia sedimentation, in detail investigated.

The capture ratio is thus computed as follows. We will choose a cylindrical co-ordinate system with the centre, which has been had in the center of the big drop, and the radial co-ordinate y, perpendicular to a heading of its falling.

let:

$$y \to y_\infty \text{ at } \tau \to -\infty.$$

Then, there is such value y_0 that at all $y_\infty \le y_0$ the small drop will face with big, and at $y_\infty \ge y_0$—will detor it. Having defined y_0, the capture factor can be computed by formula:

$$\varepsilon = y_0^2 / R_2^2. \qquad (18.48)$$

Calculations of factor of capture in oncoming purely the inertia sedimentation were for the first time Langmuir and Blodzhett(1948; 1942) are undertaken. The field of speeds of medium v in these works has been chosen as follows. For a small Reynolds number

$$\left(\text{Re} = R_2 \rho_c w_0 \left(R_0 \right) / \mu_c \right)$$

Field of speeds of Stokes, for great numbers of Rejnols the continuous potential. The analysis of usability of these models of hydrodynamic fields for calculation of factor of capture has been spent in (Rusanov, 1969; and Semrau, 1958). It has appeared that the model of a ramping flow cannot lead to a correct asymptotics at $k_c \to \infty$, and the model of a continuous potential flow leads to a little overestimated results for capture factor in all range of values k_c. For $Re \le 1$Langmuir (1948)has offered the following formulas gained by approximation:

$$\varepsilon \approx \begin{cases} \dfrac{k_c^2}{\left(k_c+0,5\right)^2} & \text{At } k_c>1/12 \text{ и } Re\gg 1 \\[3mm] \left(1+\dfrac{3}{4}\dfrac{\ln 2k_c}{k_c-1,214}\right)^{-2} & \text{At } k_c>1,214 \text{ и } Re\ll 1 \end{cases} \qquad (18.49)$$

Numerical calculations of Langmuir and Blodzhett were repeatedly mustered by other researchers (Veviorsky, 1968; and Johnstone, 1954). The divergence of results was gained small, therefore at the analysis of the inertia concretion of corpuscles of essentially different sizes, in our opinion, with sufficient accuracy, it is possible to use simple enough formulas (18.49), of course, within their usability.

For averages Re of Langmuir recommended to use interpolation formula:

$$\varepsilon \approx \frac{1}{1+Re/30}\left[\left(1+\frac{3}{4}\frac{\ln 2k_c}{k_c-1,214}\right)^{-2}+\frac{k_c^2}{\left(k_c+0,5\right)^2}\frac{Re}{30}\right]. \qquad (18.50)$$

In works (Rusanov, 1969; Kraemer, 1955; Sherwood, 1939; Fage, 1932; Lapplu, 1955; and Semrau, 1958) accuracy of this formula has been subjected carefully to the analysis. Thus for a field of speeds also were the data gained by numerical calculation of Nave–Stoks equations is taken. It has appeared that at a mean k_c and $1 \leq Re \leq 10$, the formula (18.50) leads to too underrated results. Therefore, at practical calculations of concretion of corpuscles of different sizes, we recommend to use interpolation under the data resulted on rice Figures (18.4) and (18.5).

Calculations of factor of capture have been spent at formal extending of the equation of motion of corpuscles up to its physical contact to a drop surface.

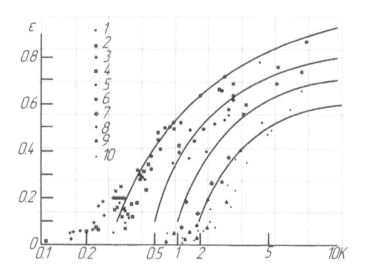

FIGURE 18.4 Dependence of factor of capture from to at a various reynolds number.

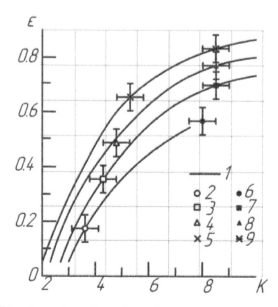

FIGURE 18.5 Comparison of experimental values of factor of capture with design data.

18.6 CONCLUSIONS

1. As appears from stated above the theory Smoluhovsky, it is fair for process of concretion of a monodisperse aerosol. In practice such aerosols meet rather seldom, the methods allowing at conservation of substantive provisions of rapid sweeping concretion to use of them for calculation of process of concretion of unequigranular aerosols therefore are necessary.

2. For calculation of process of concretion two approaches are observed. First, it is connected with correction (refinement) of formulas for definition of a constant of concretion or, more precisely, in selection of such expressions, which would allow, without penetrating into details of disperse composition of an aerosol, to gain correct values of speed of concretion. The another approach assumes working out of mathematical model of process of concretion of unequigranular aerosols.

3. The case of concretion of an aerosol in a turbulent stream, which is characterized by origination of the inertia differences between corpuscles of different sizes is observed. Owing to turbulence of a corpuscle are sped up till the various speeds depending on a size, and can then face with each other. Empirical dependences for calculation of factor of capture of corpuscles calculations on which will well be co-ordinated with experimental data are offered.

KEYWORDS

- **Capture ratio**
- **Coagulation**
- **Collisions**
- **Diffusion**
- **Monodispersity**
- **Polydispersity**

REFERENCES

1. Fuchs, N. A.; Mechanics of Aerosols. Publishing House: Academies of Sciences the USSR; **1955.**
2. Fuchs, N. A.; To the theory of a sprinkler irrigation of "warm" clouds. *The Rep Acad. Sci. USSR.***1951,***81(6),* 1043–1045 (in Russian).
3. Smoluhovsky, M.; Experience of the mathematical theory of kinetics of concretion of the colloid openings. Coagulation of Colloids.M: ONTI. **1936a,** 7–39, (in Russian).
4. Smoluhovsky, M.; Three reports on Brownian molecular motion and concretion of colloidal particles Brownian motion. M: ONTI. **1936b,** 332–417, (in Russian).
5. Green, X.; and Lein, B.; Aerosols—a Dust, Smokes and Fogs. Publishing House: "Chemistry"; **1969.**
6. Levitch, V. G.; Physical and Chemical Hydrodynamics. M: Pchisic. **1959a.**
7. The Collector "The Current State of Hydrodynamics of a Viscous Fluid." Publishing House: Lit; **1948.**
8. Fuchs, N. A.; Successes of mechanics of aerosols. In: Science Summaries, Chemical Sciences. Publishing House: Academies of Sciences the USSR; **1961,** *5.*
9. Kliachko, L. S.; Heating and Ventilation. **1931,** *4.*
10. Oseen, C.;NeuereMethoden und Ergebnisse in der Hydrodynamik. Leipzig; **1927.**
11. Chowdhury, K.; and Fritz, W.; *Chem. Eng. Sci.***1959,** *II,* 92.
12. Strauss, W.; Industrial Gas Cleaning. Pergamon Press; **1966.**
13. Van der Leeden, P.; et al. *Appl. Sci. Res.***1955,** *5A,* 338.
14. Langmuir,I.; The production of rain by chain reaction in cumulus clouds at temperatures above freezing. *J. Met.***1948,** *5,* 175–192.
15. Blodzhett, K. B.; OSRD Report No 865, **1942.**
16. Fonda, A.; and Heme, H.; Symposium on the Aerodynamic, Capture of Particles. Pergamon Press; **1960.**
17. Watson, H.; *Trans. Faraday Soc.* **1936,** *32,* 1073.
18. Levin, L. M.; Publishing House: Academies of Sciences the USSR, Sulfurs; **1957,** *7,* 914.
19. Ranz, W.; and Wong, J.; *Ind. Eng. Chem.***1952,** *44,* 1371.
20. Kousov, P. A.; Air Purification from a Dust in Cyclone Separators. **1938.**
21. Kutuzov, T. O.;Valdberg, A. J.; and Karbanov, B. N.; Industrial and Sanitary Clearing of Gases. **1971,** *2,* 16.
22. Mason, B.; Dg. Physics of Clouds. L: Hidromet Publishing House; **1961.**
23. Valdberg, A. J.; Hares, M. M.; and Padva, V. J.; Chemical and Oil Engineering Industry. **1969,** *3,* 7.
24. Valdberg, A. J.; and Larin, V. A.; Steel. **1970,** *10,* 957.
25. Valdberg, A. J.; and Fedorova, C.; Industrial and Sanitary Clearing of Gases. **1971,** *1,* 3.
26. Voloshik, V. M.; Introduction in Hydrodynamics of Coarse-Particle Aerosols. L: Hidromet Publishing House; **1971.**
27. Voloshik, V. M.; Hydrodynamic Aspects of the Theory of Concretion of Aerosol Corpuscles. The Collector Academies of Sciences; **1970,***19,* 81–115(in Russian).

28. Technical Manual:Cyclone Separators. Goshimizdat: Scientific Research Institute of Gas; **1956.**
29. Dergachev, N. F.; Fly-Ash Scrubbers of System VTI. The State Power Publishing House; **1960.**
30. Johnstone, H. F.; and Roberts, M. H.; *Ind. Eng. Chem.***1949**, *41,* 2417. Calderbank, P. H.; *Trans. Inst. Chem. Eng.***1956**, *34(1),* 79.
31. Rodions, A. I.; Kashnikov, A. M.; and Radikovsky, B. M.;.**1965**,*38(1),* 143.
32. Veviorsky, M. M.; and Dilman, V. V.; *Ind. Chem.***1968**, *7,* 1517.
33. Johnstone, H.; Field, R.; and Tassles, H.;*Ind. Eng. Chem.***1954**, *46(8),* 1601.
34. Ekman, F.; and Johnstone, H.; *Ind. Eng. Chem.***1951**, *43,* 1358.
35. Rusanov, A. A.;Urbah, I. I.; and Anastasiadi, A. P.; Clearing of smoke gases in industrial power engineering. M. "Energy." **1969.**
36. Kraemer, H. F.; and Johnstone, H. F.; *Ind. Eng. Chem.***1955**, *47(12),* 2426.
37. Sherwood, T.; and Woertz, B.; *Ind. Eng. Chem.***1939**, *31,* 1034.
38. Fage, A.; and Townend, H.; *Proc. Roy. Soc.* **1932**, *135 A,* 656.
39. Lapplu, E.; and Kamack, H.; *J. Chem. Eng. Progr.***1955**, *51,* 110.
40. Semrau, T. J.; Marynowski, C. W.; and Lunde, K. E.; *Ind. Eng. Chem.***1958**, *50,* 1615.
41. Semrau, T. J.; *Air. Poll. Control Assoc.***1960**, *10,* 200.
42. BETH-Handbuch-Staubtechnik. Lubeck: SelbstverlagMaschinenfabrik BETH Gmbh; **1964.**
43. Teverovsky, E. H.; and Hares, M. M.; Works Scientific Research Institute of Gas. **1957**, *1,* 105.
44. Ertl, D.; Stahleisen-Sonderberichte. *6(22),***1964.**
45. Deitch, M. E.; and Fillipov, G. A.; Gas kinetics of two-phase medium. M: Energy. **1981.**
46. Saltanov, G. A.; Supersonic Two-Phase Currents. Minsk: The Higher School; **1972.**
47. Hidy, J. M.; On the theory of the coagulation of noninteracting particles in Brownian motion. *J. Colloid. Sci.***1965**, *20,* 123–144.
48. Hidy, J. M.; and Brock, J. R.; The Dynamic of Aerocolloidal Systems. Oxford: Pergamon-Press; **1970.**
49. Clark, W.; and Whitby, T.; Concentration and size distribution measurements of atmospheric aerosols and a test of the theory of self-preserving size distributions. *J. Atm. Sci.***1967**,*24(6),* 677–687.
50. Coagulation of Colloids: The Collector 7. Eds. Rabinovich, A. I.; Vasileva, P. S.; M. ONTI.**1936.**
51. Tikhomirov, M. B.;Tunitskij, N. N.; and Petrjanov, I. V.; The Report. Academies of Sciences the USSR; **1942**,*94,* 865.
52. Hocking, L. M.; The collision efficiency of small drops. *Quart. J. Roy. Met. Soc.***1959**, *85(363),* 44–50.
53. Levitch, V. G.; Physical and Chemical Hydrodynamics. M: The Physical Mat Publishing House; **1959b.**
54. Herne, H.; The classical computation of the aerodynamic capture by spheres. In: Aerodynamic Capture of Particles. New York: Pergamon Press; **1960.**

CHAPTER 19

MATHEMATICAL SIMULATION OF PROCESS OF SEPARATION OF DISPERSION PARTICLES AND CHECK OF ADEQUACY OF MATHEMATICAL MODEL

CONTENTS

19.1 INTRODUCTION

One of the most promising methods for increasing the efficiency of dust collection of fine particles is scrubbing. For this method are characterized by complex mass transfer processes in the course of interaction with the gas-dispersion flow scrubbing liquid droplets, resulting in the speed and the concentration of the phase determining gas cleaning [1–4].

Existing studies in this area show a strong sensitivity of the output characteristics of the regime and design of the device, indicating that a qualitatively different flow hydrodynamics at different values of routine-design parameters [5–9].

Thus, a systematic review of the effectiveness of hydrodynamics and vortex devices, receipt, and compilation dependencies between regime-design parameters of the device, the establishment of effective and manufacturable design and development of their mass production for wide use in industrial practice is an urgent task [10–17].

19.2 DEVELOPMENT OF THE CONSTRUCTION BUBBLE-VORTEX APPARATUS WITH ADJUSTABLE BLADES

To optimize the bubble-vortex apparatus conducted experimental studies. The experiments were performed by a single method (Adler, 1986) of comparative tests on dust collector's bubble-vortex apparatus with a cylindrical chamber 0.6 m and a diameter of 0.2 and 0.4 m bubble-vortex machine with adjustable blades in accordance with Figure 19.1 comprises a cylindrical chamber 1 inlet pipe 2. The cylindrical chamber 1 is three swirl gas flows, which are four blades, curved sinusoidal curve. Adjusting the blades 2 is done by turning the eccentrics, sealed with a cylindrical chamber 1 through the spring washers and lock nuts.

Swirler before the midpoint of the gas flow nozzle 4, and after a swirler located peripheral nozzle 5, which serves scrubbing liquid. Removal of dispersed particles produced by sludge overflow pipe 6 chip catcher 7.

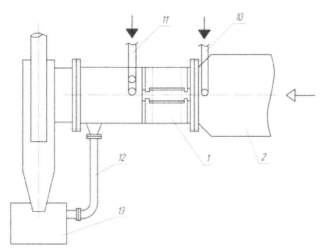

FIGURE 19.1 Research facility: (1)—cylindrical chamber, (2)—inlet pipe, (3)—swirl, (4)—central nozzle, (5)—peripheral nozzles, (6)—overflow pipe cuttings, (7)—chip catcher, and (8)—cyclone.

Bubble-vortex machine with adjustable blades works as follows. The dusty gas is fed into a cylindrical chamber 1, where the swirl 3 with blades mounted in the radial grooves of the rod, deflects the flow, and gives it a rotation. Under the action occurring at the same centrifugal forces dispersed particles are moved to the sides of the unit. For adjusting, the blades at the entrance and exit of each blade 3 have two tabs. With their help, the blade is in contact with a pair of eccentrics. Eccentrics are turning vanes on the inlet and outlet sections of the cylindrical chamber 1 in different directions. With this blade 3 are installed in the position corresponding to the most efficient gas purification.

19.3 DERIVE THE EQUATION OF MOTION OF A PARTICLE

To calculate the trajectories of the particles need to know their equations of motion. Such a problem for some particular case is solved by the author (Lagutkin, 2004).

We introduce a system of coordinates OXYZ. Its axis is directed along the OZ axis of symmetry scrubber (Figure 19.2). Law of motion of dust particles in the fixed coordinate system OXYZ can be written as follows:

$$m\frac{d\overrightarrow{v_p}}{dt} = \overrightarrow{F_{st}} \qquad (19.1)$$

where m—mass of the particle;

dv_p—velocity of the particle;

F_{st}—aerodynamic force.

For the calculations necessary to present the vector Eq. (19.1) motion in scalar form. Position of the particle will be given by its cylindrical coordinates $(r; \varphi; z)$. Velocity of a particle is defined by three components: U_p—tangential, V_p—radial, and W_p—axial velocity.

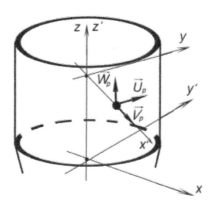

FIGURE 19.2   The velocity vector of the particle.

We take a coordinate system O'X'Y'Z ', let O'X' passes axis through the particle itself, and the axis O'Z' lies on the axis OZ. Adopted reference system moves forward along the axis OZ W_p speed and rotates around an angular velocity

$$\omega(t) = \frac{U_p}{r_p} \qquad (19.2)$$

The equation of motion of a particle of mass

$m = \frac{1}{6}\pi\rho_p d_p^3$ coordinate system O'X'Y'Z' becomes:

$$m\frac{d\vec{v_p}}{dt} = \vec{F_{st}} - m\vec{a_0} + m\left[\vec{r_p}\cdot\vec{\omega}\right] + m\left[\vec{r_p}\cdot\vec{\omega}\right] + m\left[\vec{\omega}\cdot\left[\vec{r_p}\cdot\vec{\omega}\right]\right] + 2m\left[\vec{v_p}\cdot\vec{\omega}\right]$$

or

$$\frac{d\vec{v_p}}{dt} = \frac{1}{m}\vec{F_{st}} - \vec{a_0} + \left[\vec{r_p}\cdot\vec{\omega}\right] + \left[\vec{\omega}\cdot\left[\vec{r_p}\cdot\vec{\omega}\right]\right] + 2\left[\vec{v_p}\cdot\vec{\omega}\right] \quad (19.3)$$

where a_0—translational acceleration vector of the reference frame;
dv_p—velocity of the particle;
r_p—the radius vector of the particle;

$\left[\vec{r_p}\cdot\vec{\omega}\right]$—acceleration due to unevenness of rotation;

$\left[\vec{\omega}\cdot\left[\vec{r_p}\cdot\vec{\omega}\right]\right]$—the centrifugal acceleration;

$2\left[\vec{v_p}\cdot\vec{\omega}\right]$—Coriolis acceleration.

The first term on the right-hand side of Eq. (19.3) is the force acting c
gas flow on the particle, and is given by Stokes

$$F_{st} = 3\pi\mu_g d_p\left[\vec{v_g} - \vec{v_p}\right] \quad (19.4)$$

μ_g—dynamic viscosity of the gas.
The second term Eq. (19.3) is defined as

$$\frac{dW_p}{dt}\vec{e_z} = \frac{dW_p}{dt}\vec{e_{z'}}$$

Convert the remaining terms:

$$\left[\vec{r_p}\cdot\vec{\omega}\right] = \left[\vec{r_p}\cdot\frac{d\vec{\omega}}{dt}\right] = \left[\vec{r_p}\cdot\frac{d}{dt}\left(\frac{U_p}{r_p}\vec{e_z}\right)\right] = -r_p\left(\frac{1}{r_p}\frac{dU_p}{dt} - \frac{U_p}{r_p^2}V_x\right)\vec{e_y} = \left(-\frac{dU_p}{dt} + \frac{U_pV_p}{r_p}\right)\vec{e_y}$$

$$\left[\vec{\omega}\cdot\left[\vec{r_p^{-1}}\cdot\vec{\omega}\right]\right] = \frac{U_p^2}{r_x}\left[\vec{e_z}\cdot\left[\vec{e_x}\cdot\vec{e_z}\right]\right] = -\frac{U_p^2}{r_p}\left[\vec{e_z}\cdot\vec{e_y}\right] = -\frac{U_p^2}{r_p}\vec{e_x}$$

$$2\left[\vec{v_p}\cdot\vec{\omega}\right] = 2v_{x'}\left[\vec{e_{x'}}\cdot\vec{\omega}\right] = 2v_{x'}\frac{U_x}{r_p}\left[\vec{e_{x'}}\cdot\vec{e_{z'}}\right] = \left(-2\frac{U_pV_p}{r_p}\right)\vec{e_{y'}}$$

где $\hat{e}_p$, $\hat{e}_v$, $\hat{e}_z$—vectors of the reference frame and used the fact that

$$\vec{r}_p = \vec{e}_x \cdot r_q \cdot v_x = V_p$$

Substituting these expressions in the equation of motion Eq. (19.3),

$$m\frac{d\vec{v}_p}{dt} = \vec{F}_{st} - m\vec{a}_0 + m\left[\vec{r}_p \cdot \vec{\omega}\right] + m\left[\vec{r}_p \cdot \vec{\omega}\right] + m\left[\vec{\omega}\cdot\left[\vec{r}_p \cdot \vec{\omega}\right]\right] + 2m\left[\vec{v}_p \cdot \vec{\omega}\right]$$

or

$$\frac{d\vec{v}_p}{dt} = \frac{1}{m}\vec{F}_{st} - \vec{a}_0 + \left[\vec{r}_p \cdot \vec{\omega}\right] + \left[\vec{\omega}\cdot\left[\vec{r}_p \cdot \vec{\omega}\right]\right] + 2\left[\vec{v}_p \cdot \vec{\omega}\right]$$

We write this equation in the projections on the axes of the coordinate system O'X'Y'Z'

$$\begin{cases} \dfrac{dV_{x'}}{dt} = \dfrac{1}{m}F_{stx'} + \dfrac{U_p^2}{r_p} \\[2mm] 0 = \dfrac{1}{m}F_{sty} - \dfrac{dU_p}{dt} - \dfrac{U_p V_p}{r_p} \\[2mm] 0 = \dfrac{1}{m}F_{stz} - \dfrac{dW_p}{dt} \end{cases}$$

$$\begin{cases} \dfrac{dV_p}{dt} = \dfrac{1}{m}F_{stx} + \dfrac{U_p^2}{r_p} \\[2mm] \dfrac{dU_p}{dt} = \dfrac{1}{m}F_{sty} - \dfrac{U_p V_p}{r_p} \\[2mm] \dfrac{dW_p}{dt} = \dfrac{1}{m}F_{stz} \end{cases} \qquad (19.5)$$

We have the equation of motion of a particle in a rotating gas flow projected on the axis of the cylindrical coordinate system.

Substituting Eqs. (19.2) and (19.4) in Eq. (19.5), we obtain the system of equations of motion of the particle:

$$\begin{cases} \dfrac{dV_p}{dt} = \dfrac{18\mu}{\rho_p d_p^2}\left(V_g - V_p\right) + \dfrac{U_p^2}{r_p} \\[2mm] \dfrac{dU_p}{dt} = \dfrac{18\mu}{\rho_. d_.^2}\left(U_g - U_p\right)\dfrac{U_p V_p}{r_p} \\[2mm] \dfrac{dW_p}{dt} = \dfrac{18\mu}{\rho_p d_p^2}\left(W_g - W_p\right) \end{cases} \qquad (19.6)$$

19.4 STUDY OF THE EFFECTIVENESS OF AIR CLEANING

Studies were conducted on a bubble—vortex apparatus with a cylindrical chamber with a diameter of 0.2 and 0.4 m as a model system were studied air and talcum powder with a particle size of $d = 2 \div 30$ µm. It identifies the total and fractional cleaning efficiency. We studied the effect on performance of cleaning mode parameters, which served as a—the total flow of air through the bubble—vortex apparatus, water consumption, spin factor K. Found that with increasing air flow rate is an increase cleaning, Figure 19.3.

The optimal in terms of energy limits throughput system: the lower limit corresponds to the conventional cross-sectional velocity 5 mps, the maximum flow rate is limited to a speed of 15 mps.

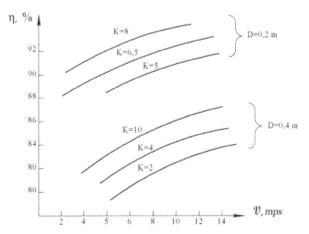

Dust: talc $d_m = 10\,mic$; $\sigma = 3,64$; $\rho_n = 2650\,{}^{kg}\!/_{m^3}$; $z = 5\,{}^{g}\!/_{m^3}$;

FIGURE 19.3 Dependence on the effectiveness of cleaning the gas flow rate.

In bubble-vortex device at the minimum speed is reduced purity, with a maximum—a sharp increase hydraulic resistance.

It has also the effect of the coefficient K spin rotator on the effective dust collection: as K increases the degree of purification. Found a range of values K, at which the relatively high trapping efficiency ($K_{min} = 5$, $K_{max} = 8$). When $K = 1$, there is a significant reduction in treatment efficiency, with $K > 10$ performance remains almost constant, but the pressure drop in the apparatus to increase substantially. The influence of the diameter of the unit to the cleaning efficiency: with increasing diameter of collection efficiency is reduced, and the smaller the median diameter of the particles, the greater the fall cleaning efficiency. The method of calculation, one with the method of calculating cyclone (Uzhov, 1975), in which the total and fractional dust collection efficiency can be determined analytically:

$$\eta = 50 \cdot \left[1 + \Phi(x')\right], \text{ where } x' = \frac{\lg\left[\dfrac{d'_{50}}{d_{50} \cdot k \cdot 10^3 \cdot \sqrt{D \cdot \dfrac{\mu_g}{\rho_p} \cdot \vartheta_g}}\right]}{\sqrt{\sigma^2 + \lg^2\left(\dfrac{d_{50}}{d_{16}}\right)}},$$

$$\eta_f = \frac{1}{2} \cdot \left[1 + \Phi(x)\right], \text{ where } x = \frac{\lg\left[\dfrac{d_\eta}{d_\eta \cdot k \cdot 10^3 \cdot \sqrt{D \cdot \dfrac{\mu_g}{\rho_p} \cdot \vartheta_g}}\right]}{\sigma},$$

where—d'_{50} the median distribution of dust particles entering the device, m; d_{50}—diameter particles captured with an efficiency of 50 per cent, m; v_g—a common velocity of the gas in the apparatus, mps; μ_g—the dynamic viscosity of gas, Pa m/s²; ρ—particle density, kg/m³; d_{16}—particle diameter at the entrance to the unit in which the mass of all the particles having a size less than 16 per cent of the total dust mass, m; σ—value that characterizes the dispersion of the particles; k—coefficient for this unit received $k = 34,75$.

When preliminary calculations, the crude treatment can be defined graphically, Figure 19.4.

C—is a function of the geometric parameters of the machine and can be calculated for vehicles designed by the known method (Leith, 1971).

Ψ—modified inertia parameter of the state of dust-gas systems.

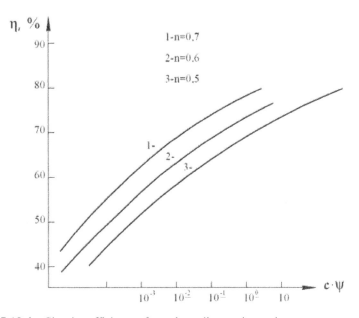

FIGURE 19.4   Cleaning efficiency of gas, depending on the product parameters $c\cdot\psi$.

$$\psi = \frac{d_p^2 \cdot \rho_p \cdot \vartheta_g}{18 \cdot \mu_g \cdot D} \cdot (n+1)$$

Factor n is given by

$$n = 1 - \left(1 - 0.0165 \cdot D^{0.15}\right) \cdot \left(\frac{T_g}{283}\right)^{0,3},$$

where T_g—the absolute temperature of the gas, K; D—diameter apparatus, m.

19.5   STUDY OF HYDRODYNAMIC CHARACTERISTICS

Investigated the resistance bubble—vortex apparatus, depending on its classification regime—design parameters.

Found that, the most efficient and cost effective mode of operation is at $K = 5 - 8$ (Usmanova, 2006).

The technique of pressure loss and specific energy consumption for dust control, which are determined by the formulas:

$$\Delta P = \xi \times \frac{\rho_g \times \vartheta^2}{2}, \, PaE = \frac{\Delta P}{3600}, kWt \times h / 1000m^3,$$

On this basis was constructed empirical mathematical model for calculating the coefficient of hydraulic resistance, including a formula to calculate ξ "dry" machine

$$\xi_{dry} = \frac{1}{n}\left((R_m)^{2n} - 1\right) + \frac{\alpha}{K^2} \cdot \left(\frac{\vartheta_{out}}{\vartheta_n}\right)^2$$

empirical relationship for calculating the pressure drop in the gas transport liquid

$$\xi_{tr} = 4 \cdot \left(\frac{Q}{G}\right)^{0,4} \cdot \sqrt{1 + \frac{1}{K^2}}$$

and ultimate dependence for calculating ξ irrigation apparatus

$$\xi_{irr} = \frac{1}{n}\left((R_m)^{2n} - 1\right) + \alpha \frac{\varepsilon^2}{K^2} \cdot \left(1 + \frac{\rho_l}{\rho_g}\right) \cdot \left(\frac{\vartheta_{out}}{\vartheta_n}\right)^2 + 4\left(\frac{Q}{G}\right)^{0,4} \cdot \sqrt{1 + \frac{1}{K^2}} \quad (19.7)$$

where ξ—coefficient of hydraulic resistance; R_m—radius of the cylindrical chamber, m; ρ_n, ρ_l—density of gas and liquid, kg/m³; v_{in} v_{ou}—velocity of gas at the inlet and outlet of the unit, mps; n, ε—indicators vortex movement, K-factor twist swirler; Q, G—liquid and gas flow kg/m³; α-twist angle of the flow,

The resulting formula takes into account the presence of the dispersed phase and the partial loss of swirling flow.

The intensity of the twisting of the gas flow ratio was estimated geometric twist K_g

$$K_g = \frac{32}{\pi^2} \cdot \frac{\vartheta_\phi}{\vartheta_x} \cdot \frac{l}{D},$$

Because the value K_g does not match the real twist coefficient, taken the following relation:

$$K = 1.4 \cdot K_g^{0.72},$$

The experimental results are presented graphically dependencies of hydraulic resistance of regime—the design parameters, Figures 19.5 and 19.6.

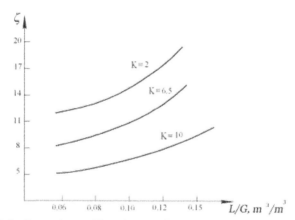

FIGURE 19.5   Dependence of ξ on the specific irrigation apparatus.

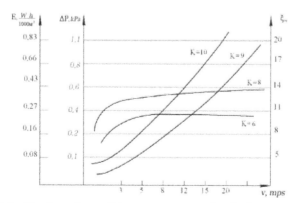

FIGURE 19.6   The dependence of the energy consumption for dust control and hydraulic resistance of the gas velocity in the machine.

19.6 OUTPUT RELATIONSHIP BETWEEN THE GEOMETRICAL AND OPERATIONAL PARAMETERS

Formal analysis of relationships that define the motion of gas and solids in the scrubber. The analysis shows that the strict observance of similarity of movements in the devices of different sizes requires the preservation of four dimensionless complexes, such as

$$\mathrm{Re}_d = \frac{wD}{v}; F_r = \frac{w^2}{Dg}; A_r = \frac{\delta \rho_2}{D \rho_1}; \mathrm{Re}_\delta = \frac{v\delta}{v}$$

Not all of these systems are affecting the motion of dust. Experimentally found that the influence of the Froude number F_r negligible (Deutsch, 1968) and can be neglected. It is also clear that the effect of the Reynolds number for large values, it is also insignificant. However, maintaining unchanged the remaining two complexes, still introduces significant difficulties in modeling devices.

On the other hand, there is no need for strict observance of the similarity in the trajectory of the particle in the apparatus. What is important is the end result—providing the necessary efficiency unit. To estimate, the parameters that characterize the removal of particles of a given diameter, consider the approximate solution of the problem of the motion of a solid particle in a scrubber. A complete solution for a special case considered in the literature (Starchenko, 1999), this solution can be used to obtain the dependence of the simplified model of the flow.

For the three coordinates—radial, tangential, and vertical equations of motion of a particle at a constant resistance can be written in the following form:

$$\frac{dw}{dt} - \frac{w_\phi^2}{r} = -\alpha\left(w_r - v_r\right)$$

$$\frac{dw_z}{dt} \cong -\alpha\left(w_z - v_z\right)$$

$$\frac{dw_\phi}{dt} + 2w_r \frac{w_\phi}{r} - \alpha\left(w_\phi - v_\phi\right)$$

where α-factor resistance to motion of a particle, divided by its mass.

$$\alpha = \frac{\mu}{K \rho \delta^2}$$

K-factor, which takes into account the effect of particle shape (take $K = 2$).

The axial component of the velocity of gas and particles are the same, as follows from the equations of motion by neglecting gravity. Indeed, if the

then taking

$$\frac{dw_z}{dt} = \alpha \left(w_z - v_z \right),$$

$$w_z - v_z = \Delta w_z,$$

we get

$$\frac{d\Delta w_z}{dt} = -\alpha \Delta w_z \quad \Delta W_z = \Delta w_{zo} e^{-dt}$$

If initially

$$w_z = v_z \quad (\Delta W_z = 0); \quad \Delta W_t = 0 \text{ и } W_t = const$$

projection speed:

$$w_\varphi = const \cdot \sqrt{r}$$

Valid law $W\varphi(r)$ may differ markedly from the accepted, but this is not essential. In this case, it only makes us enter into the calculation of average

$$\left(\frac{w_\phi^2}{r} \right)$$

Under these simplifications, the first of the equations of motion is solved in quadrature. Indeed, for now

$$\frac{dw_r}{dt} + aw_r = \frac{w_\phi^2}{r}$$

then with the obvious boundary condition t = 0, vr = 0, we have

$$w_r \cong \frac{1}{\alpha}\left(\frac{w_\phi^2}{r}\right) av \left(1 - e^{-dt}\right)$$

The time during which the flow passes from the blade to the swirler exit from the apparatus as well

$$t_1 = \frac{l}{w_{zav}}$$

On the other hand, knowing the law of the radial velocity, we can find the time during which the particle travels a distance of r_1 (the maximal distance from the wall) to the vessel wall (r_2).

$$r_2 - r_1 = \int_0 w_r dt = \frac{w_\phi}{d_r}\int_0 \left(1 - e^{-dt}\right) dt$$

$$r_2 - r_1 = \frac{v_\phi}{\alpha r_{av}}\left[t_1^l + \frac{1}{\alpha}\left(e^{-dt} - 1\right)\right]$$

Substituting in this equation is the limiting value of $t_1 = t_t$ we get:

$$\frac{\alpha r_{av}\left(r_2 - r_1\right)}{v_{\phi av}^2} \geq \frac{l}{w_z} + \frac{1}{\alpha}\left(e^{\frac{\mu}{\rho\delta^2}*\frac{l}{w_z}} - 1\right)$$

or

$$\frac{\mu}{K\rho\delta^2}\cdot\frac{r_1^2 - r_2^2}{2V_{\phi\ av}^2}\cdot\frac{w_z}{l} \geq 1 + \frac{K\rho\delta^2}{\mu}\cdot\frac{w_z}{l}\left(e^{-\frac{\mu}{\rho\delta^2 K}*\frac{l}{w_z}} - 1\right) \qquad (19.8)$$

The presented approach is based on the known dependence and model of the flow, it is different in a number of studies approach is only in the details. However, further to allocate two sets, one of which characterizes the geometry of the device, and the other—operating data. The use of these systems simplifies the calculation and, most importantly, takes into account the influence of some key factors to the desired gas velocity and height of the apparatus W_{Zav}.

Dependence structure Eq. (19.7) shows the feasibility of introducing two sets, one of which

$$A = \frac{\mu l}{2 \rho \delta^2 W_t K}$$

characterizes the effect of the flow regime and the particle diameter, and the other is a geometric characteristic of the device.

$$A_r = \frac{r_2}{l} \sqrt{\left(1 - r_1^{-2}\right)} ctg\beta \qquad (19.9)$$

In Eq. (19.8) through r_1 marked relative internal radius apparatus:

$$r_1 = r_1 / r_2$$

and β_1—the average angle of the flow at the exit of the guide apparatus

$$tg\beta_1 = \frac{V_\phi}{V_z}$$

Then Eq. (19.7) takes the simple form

$$A_r \geq f(A)$$

where

$$f(A) = \sqrt{\frac{1}{A} + \frac{1}{2A^2}\left(l^{-2A} - 1\right)} \qquad (19.10)$$

One of the important consequences of the resulting function is the relationship between the diameter of the dust particles and the axial velocity of the gas. With this machine, with data A_g, $A_{min} = const$ and therefore

$$w_z \delta^2 = const$$

This means that reducing the particle size axis (expenditure), the rate should be increased according to the dependence

$$\frac{w_z}{w_{zo}} = \left(\frac{\delta_o}{\delta}\right)^2$$

Unfortunately, a significant increase W_z permitted as this may lead to the capture of dust from the walls and ash. You can also change the twist angle β and the height of the flow system, without changing the axial velocity.

If the reduction of the particle diameter δ or increasing the size of the unit has increased the value of A, it must be modified accordingly A_g (using a graph Figure 19.3), and the new value of A_r to find an angle β:

$$\frac{ctg\beta}{ctg\beta_0} = \left(\frac{A_g}{A_{go}}\right)$$

19.7   CALCULATING THE COST OF DUST CONTROL

Using previously obtained correlations between the dust collection efficiency and pressure drop with regime—the design parameters of design procedure. The technique allows to select the unit with operational and design parameters under which it would provide the required process parameters at a minimum cost of cleaning.

The main components of the cost of treatment are worth the dust is not trapped unit (C_n), and the cost of energy in transmission through the apparatus of the gas flow (C_e).

The cost of treatment is determined by the formula:

$$C_0 = C_n + C_e,$$

Cost crude dust C_n decreases with increasing the efficiency of the device, with a decrease in the initial concentration of the dust, and with a decrease in the cost of the collected dust C_u:

$$C_n = (1-\eta) \cdot z_1 \cdot C_u,$$

The cost of electricity consumed for dust control, increases with increasing hydraulic resistance unit and is calculated by the formula:

$$C_e = \Delta P \cdot Q \cdot r \cdot C_u$$

The full expression for calculating the cost of cleaning a gas cubic meter can be obtained using the formula (19.1) to calculate the efficiency and Eq. (19.2) to calculate the hydraulic resistance.

19.8 CONCLUSIONS

1. Developed a method of calculating the total and fractional dust collection efficiency, taking into account the geometric parameters of the device.
2. Developed a method for calculating the hydraulic resistance of a bubble—a vortex system, taking into account the structural parameters of the swirler and the presence of the dispersed phase.
3. The resulting formulas were the basis for the development of methods for calculating bubble—vortex apparatus. The developed method allows to calculate the optimum geometry devices working optimally. Optimization criterion is the minimum cost of cleaning unit volume gas along with the required efficiency of dust collection.
4. The developed method can be used in the calculation and design of gas cleaning devices, as constituent relations define the relationship between the technological characteristics of the dust collectors and their geometrical and operational parameters.
5. Creating a mathematical model of the motion of a particle of dust in the swirling flow allowed us to estimate the influence of various factors on the collection efficiency of dust in the offices of the centrifugal type, and create a methodology to assess the effectiveness of scrubber.
6. Identified settlement complexes, one of which characterizes the geometry of the scrubber and the other operational parameters. The use of these systems and simplifies the calculation takes into account the influence of several key factors.
7. The developed method can be used in the calculation and design of gas cleaning devices, as constituent relations define the relationship between the technological characteristics of the dust collectors and their geometrical and operational parameters

KEYWORDS

- **Cost of gas purification**
- **Custodial complex**
- **Efficiency gas purification**
- **Geometric complex**
- **Hydraulic resistance**
- **The coefficient of resistance**
- **The trajectory of particles**
- **Vortex apparatus**

REFERENCES

1. R. R. Usmanova. Patent 2234358 *RF bubble-vortex machine with adjustable blades* Bull. No 23. **2004**.
2. Adler, J. P.; Experimental Design in the Search for Optimal Conditions. Eds. Adler, J. P.; Markov, E. V.; and Granovsky, Y.; Moscow: Nauka; **1986.**
3. Uzhov, V. N.; Preparation of Industrial Gases to Clean. Uzhov, V. N.; Valdberg, A.; M.: Chemistry; **1975.**
4. Leith, D.; Aiche, Symposium Series. Leith, D.; and Licht, W.; Air. **1971.**
5. Usmanova, R. R.; Complex aerohydrodynamic research and the efficiency of arresting particles for barbotage—rotation. Usmanova, R. R.; Panov, A. K.; Zaikov, G. E.; *J. Balkan Tribol. Assoc.* **2006**, *3*, 368–373, *(in Russian).*
6. Barahtenko, G. M.; and Idelchik, I. E.; Industrial and Sanitation Gas. **1974**, *6*, 4–7, *(in Russian).*
7. Straus, V.; Industrial Cleaning Gases. Moscow: Chemistry; **1981.**
8. Shilyaev, M. I.; Aerodynamics and Heat and Mass Transfer of Gas-Dispersion Flow: Studies Allowance. Tomsk: Publishing House of Tomsk. State Architect Builds University; **2003.**
9. Lagutkin, M. G.; Tohti Sheep. **2004**, *38, 1,* 9–13, *(in Russian).*
10. Deutsch, M. E.; and Filippov, G. A.; Gas Dynamics of Two-Phase Media. Moscow: Energy; **1968.**
11. Starchenko, A. V.; Bells, A. M.; and Burlutskiy, E. S.; Thermophysics and Aeromechanics. **1999**. *6, 1,* 59–70, *(in Russian).*
12. Gupta, A.; Lilly, D.; and Sayred, N.; Swirling Flow Trans. from English. Ed. Krashennikova, S. Y.; Mir Moscow; **1987**.
13. Bezic, D. A.; Diss. Kand. Tehn. Sciences, Gosinzh-Tehnol. Akademiya Bryansk, Bryansk; **2000.**
14. Mizonov, V. E.; Blaschek, V.; and Colin, R.; Greeks Tohti. **1994**, *28(3),* 277–280, *(in Russian).*

15. Litvinov, A. T.; *J. Appl. Chem.* **1971,** *44(6),* 1221–1231, (*in Russian*).
16. Kutepov, A. M.; Hydrodynamics and Heat Exchange at Steam Formation. "Higher school"; **1977.**
17. Jofinov, G. A.; Collection: Institutes of a Lab Our Safety of the All-Union Central Council of Trade Unions. **1975.**

CHAPTER 20

MODERNIZING AND COMMERCIAL OPERATION OF INSTALLATIONS FOR REFINING OF GAS EMISSIONS

CONTENTS

20.1   INTRODUCTION

The Analysis of merits and demerits of various methods of clearing of gas from gaseous impurity, such as condensation, adsorptive, absorpritive a method, has allowed to draw a conclusion that at clearing great volumes of gas emissions by the most simple and reliable in realization is absorpritive a method. However, traditionally applied absorpritive the equipment packing, bubbling, and atomization types possesses low throughput on gas, and, hence, in greater{big} overall dimensions[1-3].

In this connection, use of devices of vortical type is in some cases unique by the decision of a problem.

Comparison of opportunities dust separation the equipment of various types also testifies to prefer ability of application of vortical devices at clearing great volumes of gas emissions.   All   above-stated   has served for development of designs of vortical devices of new generation.

20.2   BUBBLING—THE VORTICAL DEVICE

The Optimum mode of clearing of gas emissions can be reached{achieved} at application of the developed design bubbling—the vortical device providing a mode evaporation—condensation cooling and centrifugal wet dust-separation[4-7].

The developed design bubbling—the vortical device is presented in Figure 20.1—bubbling—the vortical device (longitudinal section). Bubbling—the vortical device according to Figure 20.1 contains the cylindrical chamber 1 with an entrance pipe. In the cylindrical chamber 1, it is established{installed} vortex generator 2 gas streams, representing pair the crossed planes forming four blades, forming flowing section. In bubbling—the vortical device before vortex generator a gas stream 2 the central atomizer is established{installed}, and in each flowing section after vortex generator 2 peripheral atomizers 3 are located.

Tap{Removal} of disperse particles from the cylindrical chamber 1 is carried out by means of a pipe of an overflow slime in slime collector.

Bubbling—the vortical device works as follows. Dusty gas moves in the cylindrical chamber 1 on an entrance pipe where vortex generator 2 by means of the blades forming flowing section, rejects a stream and gives to it{him} rotary movement. Under action of centrifugal force arising at it{this} disperse particles move to walls of the cylindrical chamber 1. For improvement of conditions of clearing of gases, before and after vortex generator 2 are established{installed} one central and four peripheral 3 atomizers in which the irrigating liquid moves. The central atomizer established{installed} before vortex generator 2, creates a volumetric torch atomization an irrigating liquid. At contact of the polluted gas and a liquid, there is a partial evaporation of last and cooling of gas. The formed suspension is divided{shared} under action of the centrifugal force arising at rotation of a stream. The torch atomization the cooling liquid, formed by the central atomizer (alongside with action of centrifugal forces), promotes outflow of disperse particles from the central zone of the cylindrical chamber 1 that reduces a way of a particle up to a wall and reduces time of separation.

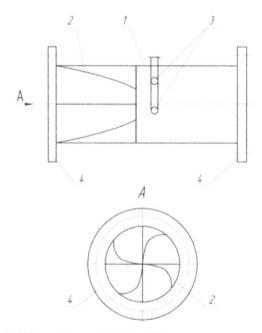

FIGURE 20.1 Bubbling—the vortical device [1].

The Optimum geometry of blades vortex generator a gas stream is provided with that on an entrance site of the blade surfaces of the rectangular flat form have, to a target site passing in parabolic. Thus, blades bending and the greater party{side} concern{touch} an internal surface of the cylindrical chamber 1 and are rigidly attached to it{her} on all to their length.

Separated slime, it is washed off by a liquid, sprayed by four peripheral atomizers 3 established{installed} in each flowing section after vortex generator 2, and by means of an inclination of the cylindrical chamber 1, it is transported on a pipe of an overflow slime in slime collector. The corner of an inclination of the cylindrical chamber 1 is picked up by practical consideration within the limits of 5–8, sufficient for tap{removal} slime.

At contact of gas suspension with a cold irrigating liquid, there is a further cooling cleared gas and integration of not separated disperse particles due to condensation water pair on the last.

Technical and economic efficiency of use offered bubbling—the vortical device for clearing and cooling of smoke gases consists that it{he} allows:

1. To raise{increase} efficiency dust separation due to installation in the device vortex generator a gas stream, the central, and peripheral atomizers;
2. To lower hydraulic resistance of the device owing to a choice of optimum geometry of blades vortex generator;
3. To save material means and the areas of industrial premises{rooms} due to an opportunity of installation bubbling—the vortical device in vent dust removal system.

On the developed design bubbling—the vortical device for clearing and cooling of smoke gases the Patent of the Russian Federation No 2182843 is received [1].

20.3   BUBBLING—THE VORTICAL DEVICE WITH ADJUSTABLE BLADES

Bubbling—the vortical device with adjustable blades according to Figure 20.2 contains the cylindrical chamber 1 with an entrance pipe 2. In the cylindrical chamber 1, it is established{installed} vortex generator 3 gas streams, representing four blade 4, bending according to Figure20.3 on a sine wave curve and having on entrance and input section rectangular ledges 3 by means of which blades are fixed in radial grooves of the core 4 placed on an axis of the cylindrical chamber 1 and rigidly connected with it{her} carving stud 5, screwed in cylindrical chamber 1 with additional welded thickenings. On an input{entrance} and an output{exit} of each blade 2 ledges 6 by means of which the blade is in contact to pair clowns 7 are stipulated. Adjustment of position of blades 2 is carried out by turn of the clowns 7 fastened to the cylindrical chamber 1 by means of spring washers 8 and counter nuts 9.

Before vortex generator a gas stream the central atomizer 10 is established{installed}, and after vortex generator peripheral atomizers 11 in which the irrigating liquid moves are located. Tap of disperse particles from the cylindrical chamber 1 is carried out by means of a pipe of an overflow slime 12 in slime collector 13.

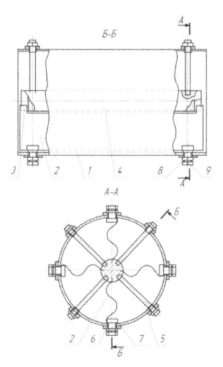

FIGURE 20.2   Bubbling—the vortical device with adjustable blades [2].

Bubbling—the vortical device with adjustable blades works as follows.

Dusty gas moves in the cylindrical chamber 1 where vortex generator by means of the blades 2 fixed in radial grooves of a core 4, rejects a stream and gives to it{him} rotary movement. Under action of centrifugal force arising at it{this} disperse particles move to walls of the device. For regulation of position of blades 2 on an input{entrance} and an output{exit} of each blade 2 two ledges by means of which the blade is in contact to pair clowns 7 are stipulated. Clowns 7 carry out turn of blades on entrance and target sites of the cylindrical chamber 1 in various directions owing to what blades 2 are established{installed} in the position corresponding{meeting} the greatest efficiency dust separation.

If necessary blades 2 come back in initial position due to elasticity of a material of blades. After adjustment andachievements of the greatest efficiency of clearing clowns 7 are fixed in this position by means of spring washers and counter nuts. Blades vortex generator bundling on a sine wave curve that promotes decrease{reduction} in hydraulic resistance of the device.

For improvement of conditions of clearing up to vortex generator one central atomizer 10 is established{installed}, the irrigation from which promotes outflow of disperse particles from the central zone of the cylindrical chamber 1 that reduces a way of particles up to a wall and reduces time of separation.

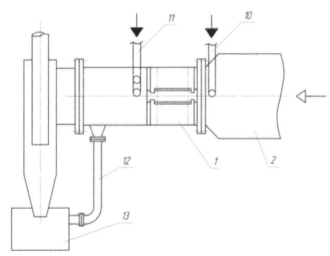

FIGURE 20.3   Installation for clearing and cooling of gas emissions.

Separated slime it is washed off by a liquid, sprayed by four peripheral atomizers and by means of an inclination of the cylindrical chamber 1 it is transported on a pipe of an overflow slime 21 in slime collector 13, according to Figure 20.3

20.4   ROTOKLON WITH ADJUSTABLE SINE WAVE BLADES

Cylindrical vortex generator with the central input of a stream within the limits of at $z=0$ represents nozzles. It{he} has found application for catching drops and distributions of ventilating air in premises{rooms} [4].

The Invention concerns to devices for wet clearing gases and application in various industries where clearing of gases of the weighed impurity is required can find.

Increase of efficiency dust separation is caused by an opportunity of regulation of position of the bottom blades concerning top that allows creating an optimum operating mode of the device with high efficiency dust separation in a wide range of a dust content of a gas stream.

The more a corner of a deviation, the above intensity of coagulation and efficiency of clearing of gas.

Rotoklon, it is characterized by presence of several slot-hole channels, form the top and bottom blades, and in each subsequent on a course of gas the channel the bottom blade is established{installed} above previous.

Such arrangement promotes a gradual input {entrance} water gas a stream in slot-hole channels and reduces that hydraulic resistance of the device.

The Arrangement of an entrance part of blades on an axis with an opportunity of their turn allows creating an active zone of diffusion. Consistently located slot-hole channels create in diffusion to a zone formed by a corner of turn of blades, a hydrodynamical zone of intensive wetting of particles of a dust. In process of moving a stream through a liquid veil, the opportunity of repeated stay of particles of a dust in hydrodinamic to an active zone is provided that considerably raises{increases} efficiency dust separation and ensures the functioning into the device in wide ranges of a dust content of a gas stream.

In Figure 20.4, longitudinal and cross-section cuts{sections} rotoklon are presented.

Rotoklon with adjustable sine wave blades, according to Figure 20.1 contains the case 1 with branch pipes for an input{entrance} 2 and an output{exit} 3 gases in which pairs blades of a sine wave structure are established{installed}. Moving of the top blades 4 is carried out by means of screw lifts 5, the bottom blades 6 are fixed on an axis 7 with an opportunity of their turn. The corner of turn of the bottom blades 6 gets out of a condition of constancy—speeds dust gas a stream. For regulation of a corner of turn of a target part of the bottom blades 6 flywheels 8 are stipulated.

The Quantity {Amount} of pairs blades is defined {determined} by productivity of the device and a dust content dust gas a stream that is a mode of steady work of the device.

In the bottom part of the case, there is a branch pipe for plum slime waters 9. Before a branch pipe for an output{exit} of gas 3, it is established{installed} labyrinth to a drop-catcher 10.

Rotoklon works as follows.

Depending on a dust content dust gas a stream the top blades 4 by means of screw lifts 5, and the bottom blades 6 by means of flywheels 8 are established{installed} on a corner defined{determined} by an operating mode of the device.

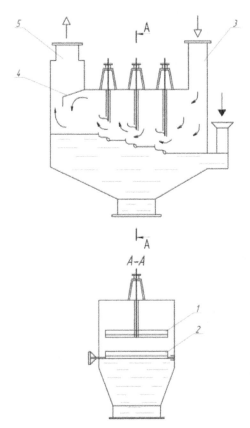

FIGURE 20.4   Rotoklon with adjustable sine wave blades [4].

Dusty gas acts in an entrance branch pipe 2 in the top part of the case of 1 device. Hitting about a surface of a liquid, it{he} changes the direction and passes{takes place} in the slot-hole channel formed top 4 and bottom 6 blades. Owing to high speed of the movement, cleared gas grasps the top layer of a liquid and splits up it{him} in the smallest drops and foam with an advanced surface.

After consecutive passage of all slot-hole channels gas passes{takes place} through labyrinth drop catcher 10 and through a target branch pipe 3 leaves in an atmosphere. The caught dust settles in the bunker to rotoklon and through a branch pipe for plum slime waters 9, together with a liquid, is periodically deduced{removed} from the device.

20.5   BUBBLING—THE VORTICAL DEVICE WITH SCREW VORTEX GENERATOR

The Invention is directed on an intensification of process of clearing of gas emissions due to uniform distribution of a liquid in a gas stream and an effective utilization of working volume of the device.

Increase of efficiency dust separation is caused by installation in the device of the axial sprinkler punched by apertures of small diameter, placed on a screw line. Apertures are executed in such a manner that the jet of a liquid from them is directed tangential to a surface of an axial sprinkler. The corner of an inclination of a jet is provided with directing plugs, compaction in apertures of an axial sprinkler. Thus due to impact following under a pressure the spiral film-drop liquid veil is formed of apertures of an axial sprinkler of liquid jets with a peripheral edge of blades vortex generator.

Blades vortex generator, executed in the form of screw, are established{installed} on all length separation zones of the device that enables to support{maintain} intensity twist a constant on length separation zones at $\beta_1 = \beta_2$ or to change it{her} under in advance chosen law at $\beta_1 \neq \beta_2$.

Bubbling—the vortical device with screw vortex generator works as follows.

Dusty gas moves in the cylindrical chamber 1 on an entrance pipe 2 where vortex generator 6 gas streams by means of the blades executed in the form of screw, rejects a stream and gives to it{him} rotary movement.

Under action arising thus of centrifugal force disperse particles move to walls of the cylindrical chamber 1.

For improvement of conditions of clearing of gas on length of the cylindrical chamber 1 (Figure 20.5) the axial sprinkler 3 punched—is established{installed} by apertures of small diameter 4 in which the irrigating liquid moves.

The Jet of a liquid from apertures 4, is directed tangential to a surface of an axial sprinkler 3 due to installation of directing plugs 5, thus the spiral film-drop liquid veil due to impact of a jet of a liquid with a peripheral edge of blades vortex generator 6 is formed. At contact of gases and a liquid, there is its{her} partial evaporation and cooling of gas. The formed suspension is divided{shared} under action of the centrifugal force arising at rotation of a stream. Separated slime by means of an inclination of the cylindrical chamber 1, it is transported on a pipe of overflow slime 10 in slime collector 11. The subsequent division of suspension occurs{happens} in a cyclone 8 attached to the cylindrical chamber 1 by means of flanges 9, whence slime also acts in slime collector 11 (Figure 20.6).

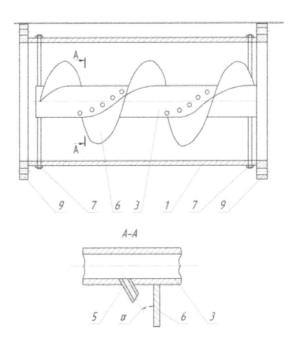

FIGURE 20.5   The longitudinal section of the cylindrical chamber.

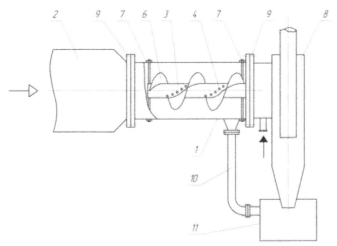

FIGURE 20.6 The general view bubbling—the vortical device with screw vortex generator.

20.6 BUBBLING—THE VORTICAL DEVICE WITH AN AXIAL SPRINKLER

In the developed device clearing of gas emissions is carried out due to more effective utilization of working volume of the device that intensifies process of gas purification.

Increase of efficiency dust separation is caused by installation in the device of the axial sprinkler representing an axial branch pipe, muffled with the target end and punched on all length apertures of small diameter on which length blades vortex generator are rigidly attached, forming flowing section. Owing to that apertures are placed on length of the cylindrical chamber, in a zone of submission of dusty gas the surface contact phases increases and, more the effective utilization of working volume of the device hence takes place.

The Sprinkler fastens to the cylindrical chamber carving stud, screwed in radial carving apertures in a sprinkler on the one hand and connected with the cylindrical chamber with another, connection is fixed by means of counter nuts.

The Device works as follows:

Dusty gas moves in the cylindrical chamber 1 on an entrance pipe 2 where vortex generator 6 by means of the blades forming flowing section

7, rejects a stream and gives to it{him} rotary movement. Under action of centrifugal force arising at it{this} disperse particles moveto walls of the cylindrical chamber 1.

For improvement of conditions of clearing in the cylindrical chamber 1 the axial sprinkler 3 in which the irrigating liquid moves is established {installed}. Owing to that the sprinkler 3 is punched by the apertures of small diameter 4 located in each flowing section 7 vortex generator 6, the surface of contact of phases in consequence{investigation} dispersion liquids on all volume of the cylindrical chamber 1 increases. At contact of gases and a liquid, there is a partial evaporation of last and cooling of gas. The formed suspension is divided {shared} under action of the centrifugal force arising at rotation of a stream. Dispersion of an irrigating liquid promotes outflow of disperse particles from the central zone of the cylindrical chamber 1 that reduces time, separation. The optimum geometry of blades vortex generator 6 is provided with that blades have a surface of the hyperbolic form, allowing to lower hydraulic resistance. Thus blades are rigidly fastened to an axial sprinkler 3, forming flowing section 7 (Figure 20.7).

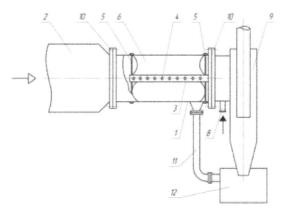

FIGURE 20.7   The general view bubbling—the vortical device with an axial sprinkler.

Separated slime, it is washed off by a liquid and by means of an inclination of the cylindrical chamber 1, it is transported on a pipe of an overflow slime 11 in slime collector 12, the Corner of an inclination of the cylindrical chamber 1 is picked up by practical consideration within the limits of 100, sufficient for tap{removal} slime. The subsequent division of suspension occurs{happens} in a cyclone 9, whence slime also acts in slime collector 12 (Figure 20.8).

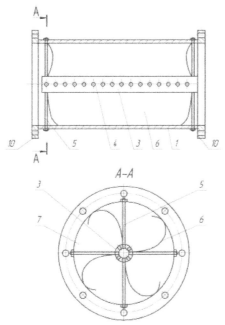

FIGURE 20.8   Longitudinal and cross-section cuts{sections} of the cylindrical chamber [3].

20.7   BUBBLING—VORTICAL GAS WASHER

The Technical result provided bubbling—vortical gas washer is expressed in increase of efficiency dust separation due to installation in the device of the axial branch pipe punched by apertures of small diameter, and having on an entrance site conic fairing, from the opposite party{side}—conic a reflector.

Increase of efficiency dust separation is caused by installation in the device vortex generator in the form of the blades forming flowing section. Blades, it is rigid put on the axial branch pipe punched by apertures for submission of the irrigating liquid. Owing to that apertures are placed on length of the cylindrical chamber, in a zone of submission of dusty gas the surface contact phases increases and, hence, the effective utilization of working volume of the device takes place more.

Decrease{Reduction} in power inputs of the device is caused by action conic fairing, which provides smooth passage of a gas stream through vortex generator that reduces the hydraulic losses characterized in factor of hydraulic resistance.

Owing to that the offered device has low factor of hydraulic resistance, there is an opportunity to raise{increase} speed of a gas stream that provides more intensive display centrifugal—inertial forces and raises{increases} efficiency of process of tap{removal} of firm particles on walls of the cylindrical chamber. The firm particles which have been not caught by a drop liquid under action of centrifugal forces, also change a direction of the movement and direct to an internal surface of the chamber covered by a film of a liquid, and are grasped by her{it}.

Owing to presence of a reflector the particles, which have been not deduced{removed} from a gas stream under action of centrifugal forces, again are rejected on walls of the chamber. Connection of the cylindrical chamber with an inclination to a cyclone in view of a drain slime provides tap{removal} of disperse particles in slime collector that also raises{increases} efficiency dust separation.

Besides bubbling—vortical gas washer can be used as the independent dust removal device, and to be established{installed} in vent dust removal system with the purpose of economy of material means and the areas of industrial premises{rooms}.

In Figure 20.9 longitudinal and cross-section cuts{sections} of the cylindrical chamber are presented

Bubbling—vortical gas washer according to Figure 20.9 contains the cylindrical chamber 1 with an entrance pipe 2. In the cylindrical chamber, it is established {installed} vortex generator 3, representing four blade, it is rigid put on the axial branch pipe 4 punched by apertures of small diameter 5. The entrance site of a branch pipe 4 is supplied conic fairing 6, and the opposite end—by a reflector 7 in the form of the truncated return cone. The liquid on irrigation moves in an axial branch pipe 4 on a pipe 8. The cyclone 9 is connected to the cylindrical chamber 1 by means of connecting flanges 10. Tap{removal} of disperse particles is carried out from the cylindrical chamber by means of a pipe of an overflow slime 11 in slime collector 12.

Bubbling—vortical gas washer works as follows:

Dusty gas moves in the cylindrical chamber 1 on an entrance pipe 2 where vortex generator 3 by means of the blades forming flowing section, rejects a stream and gives to it{him} rotary movement. Under action of centrifugal force arising at it{this} disperse particles move to walls of the cylindrical chamber 1.

For improvement of conditions of clearing in the cylindrical chamber 1, the axial branch pipe 4 in which the irrigating liquid moves is established{installed}. Owing to that the axial branch pipe 4 is punched by apertures small, diameter 5, located in each flowing section vortex generator 3, the surface of contact of phases owing to dispersion liquids on all volume of the cylindrical chamber 1 increases. At contact of gases and a liquid, there is a partial evaporation of last and cooling of gas. The formed suspension is divided{shared} under action of the centrifugal force arising at rotation of a stream. Dispersion of an irrigating liquid promotes outflow of disperse particles from the central zone of the cylindrical chamber 1 that reduces time of separation. Conic fairing 6 provides smooth passage of a gas stream through vortex generator 3. The reflector 7 promotes uniform distribution of a film of a liquid on an internal surface of the cylindrical chamber 1 that considerably raises{increases} a surface of contact of phases and increases the area of a zone of capture of the firm particles rejected on walls of the cylindrical chamber by 1 film of a liquid. The degree of clearing of gas as a result raises{increases}.

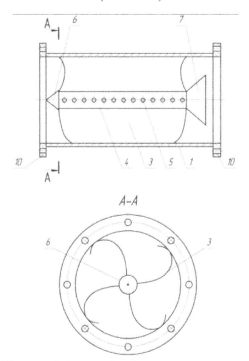

FIGURE 20.9 Bubbling—yortical gas washer [3].

20.8 ROTOR BUBBLING—THE VORTICAL DEVICE

The Invention is directed on an intensification of process of gas purification due to more effective utilization of action of centrifugal forces and increases in a surface of contact of phases.

Increase of efficiency dust separation is caused by tangential input of a gas stream, which under action of a rotating rotor gets rotary screw movement and simultaneously processed by irrigating liquid. Thus, it is formed spiral film—a drop liquid veil due to impact of a jet of the irrigating liquid sprayed from atomizers, with edges of blades screw. Blades screw are established{installed} on all length separation zones of the device that enables to support{maintain} intensity of action of centrifugal forces on length separation zones.

Efficiency of clearing is reached{achieved} also owing to that screw is executed in the form of a brush from polyamide strings. Blades of a screw brush concern{touch} with the free ends internal wall the cylindrical chamber. Rejected on walls of the cylindrical chamber slim (deposit) at once is cleaned off with wall and transported by a rotating screw brush to a pipe of an overflow slim.

Connection of the cylindrical chamber with an inclination to a cyclone in view of a drain slim also provides tap{removal} atomizer particles in slim collector.

In Figure20.10 the longitudinal section of the cylindrical chamber is presented.

Rotor bubbling—the vortical device according contains the cylindrical chamber 1 with a tangential branch pipe 2 for input of cleared gas. Inside of the chamber on two support, the rotor 3 on which it is established{installed} screw 4, executed in the form of a brush from polyamide strings is placed. On an input{entrance} in the cylindrical chamber 1 the atomizers 5 directed towards to a stream of gas are established{installed}. The cylindrical chamber is attached with an inclination to a cyclone 6. Tap{removal} of disperse particles is carried out on a pipe of an overflow slim 7 in slim collector 8.

Rotor bubbling—the vortical device works as follows. Dusty gas moves in the cylindrical chamber 1 on a tangential branch pipe 2 where under action of a rotating rotor 3 and screw 4 gets rotary screw movement. Simultaneously gas stream is irrigated with a liquid acting from atomizers 5. Jets of the liquid sprayed from atomizers 5, form a liquid veil, which provides repeated change of an interphase surface. Rejected on walls of the cylindrical chamber 1 шлам (deposit) is cleaned off and transported screw 4 to a pipe of an overflow шлама 7. The subsequent division of suspension occurs{happens} in a cyclone 6, whence slime acts in slime collector 8.

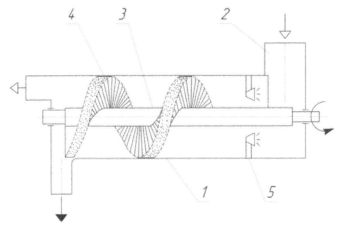

FIGURE 20.10   The longitudinal section rotor bubbling—the vortical device.

20.9   MAGNETIC HYDROCATCHER

The Invention concerns to devices for wet clearing gases and application in various industries where clearing of gases of the weighed impurity is required can find.

Increase of efficiency of clearing of gas is caused by action on a gas stream of two forces promoting dust separation: the forces of inertia arising at progress of a gas stream, and the centrifugal force arising at rotation of a stream.

Decrease{Reduction} in the specific charge of a dust removal liquid occurs{happens} owing to its{her} rotary movement together with a magnetic liquid due to action of forces of hydrodynamical friction. Thus, the surface of a liquid accepts the funneled form that allows to increase a surface of contact of a dust removal liquid with gas without increase in its {her} volume and, hence, to raise{increase} efficiency of clearing of gas.

Accommodation of a drain branch pipe on the center of the bottom promotes the most effective removal{distance} slime, concentrating under action of hydrodynamical forces at an axis of rotation of a stream.

In Figure 20.11, the magnetic hydrocatcher, a longitudinal section is presented.

The Magnetic hydrocatcher according to Figure 20.11 contains confuser 1 with a branch pipe 2 inputs of gas in which the atomizer 3 preliminary irrigation and maintenance of the set level of a dust removal liquid 4 in the case 5 is established{installed}. Coaxially cofuser 1, it is located diffuser 6. On the center of the bottom of the case 5, the drain branch pipe 7 for conclusion slime is placed. The electromagnetic system includes a magnetic liquid 8, a restrictive ring 9 and magnetic coils with a winding 10.

The Magnetic hydrocatcher works as follows.

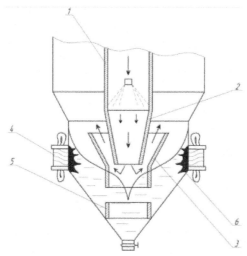

FIGURE 20.11 The magnetic hydro catcher.

In working order, the case 5 is filled with a dust removal liquid 4 up to a level of the bottom basis confuser 1. Further the electromagnetic system is connected, in magnetic coils 10 the variable pressure{voltage} is created, thus inside of the case 5, there is a rotating magnetic field. The magnetic liquid 8 is drawn by this field to an internal wall of the case and also resulted{brought} in rotation. The dust removal liquid 4, owing to forces of hydrodynamical friction, starts to rotate together with a magnetic liquid 8, forming funnel on the surface.

The Dusty gas stream hits about a surface of a liquid 4 and changes the direction. Thus, under action of centrifugal forces, it{he} also is resulted{brought} in rotary movement. Rotating water gas the mix passes{takes place} between walls confuser 1 and diffuser 6, thus the effect of pipe Venturi is formed.

Besides under action of centrifugal forces, particles of a dust are rejected to walls of the case 5 that increases time of their interaction with a liquid 4 and raises{increases} efficiency dust separation. Tap{removal} slime is carried out on the drain branch pipe 7 placed in the center of the bottom that promotes the fullest removal{distance} slime at the minimal charge of a dust removal liquid 4.

20.10 DYNAMIC GAS WASHER

The Invention concerns to devices for wet clearing gases and application in various industries where clearing of gases

The Invention is directed on increase of efficiency of clearing of gas from mechanical and gaseous impurity due to more effective utilization of action of centrifugal forces and increase in a surface of contact of phases.

Increase of efficiency of gas purification is reached{achieved} due to installation conic dissector promoting formation{education} of flat radial jets of a liquid which, being reflected from the case of the device create a high-intensity surface of contact of phases.

The Centrifugal forces arising at rotation of a rotor provide crushing a liquid on fine drops that causes intensive contact of gases and caught particles to a liquid.

Owing to action of centrifugal forces, intensive hashing of gas and a liquid and presence of the big interphase surface of contact, there is an effective clearing of gas in a foamy layer.

Dynamic gas washer, according to Figures 20.12 and 20.13, contains the vertical cylindrical case 1 with the bunker 2 gathering slime, branch pipes of input{entrance} 3 and an output{exit} 4 gas streams.

Inside of the case, it is established{installed} conic vortex generator 5, containing

The ring top basis 6 and conic dissector 7, placed on the bottom basis 8 vortex generator 5. The bases 6 and 8 are rigidly connected with each other by means of the unidirectional blades 9.

Before vortex generator 5, the axial branch pipe 10 supplies of a liquid is placed, at a level top cut, which the basis conic dissector 7 is located with a backlash. On an output{exit} of a gas stream are stipulated drip pan 11. Vortex generator 5, it is fixed on a rotor 12, which is resulted{brought} in rotation by the electric motor through a belt drive 13.

Dynamic gas washer works as follows.

The Gas stream containing mechanical or gaseous impurity, acts on a tangential branch pipe 3 in the ring space formed by the case 1 and rotor 12. The liquid acts in the device by means of an axial branch pipe 10. Owing to that the axial branch pipe 10 supplies of a liquid is placed before vortex generator 5, at dispersion liquids the zone of contact of phases increases and, hence, the effective utilization of working volume of the device takes place more.

Accommodation of the basis conic dissector 7 at a level top cut an axial branch pipe 10 provides submission of a liquid in the device in the form of the flat radial jets, which are spitted up about walls of the case 1 and forming liquid veil.

The Ring top basis 6 filling free space, formed at rotation water gas a layer, promotes stabilization water gas a stream and interferes overshoot the polluted gas.

The Gas stream, passing{taking place} rotating vortex generator 5, gets the screw movement characterized by intensive turbulent pulsations in a kernel of a stream.

At rotation of a rotor 12, liquid takes a great interest in blades vortex generator 5, moistens their surface and is sprayed in the form of fine drops on a way of movement of cleared gases that provides high-intensity contact of phases outside of conic vortex generator 5 and raises{increases} efficiency of clearing of gas.

Twirled water gas, the mix cooperates with blades 9, is intensively spitted up that leads to updating of a surface of contact of phases and an intensification of processes of an interphase exchange.

Caught atomizer particles in the form of slime continuously leave from the bunker 2.

The gas Cleared of impurity together with sparks of a liquid passes{takes place} through drip pan 11, is released{exempted} from a part splashes and, bending around a branch pipe 10, leaves{abandons} the device through a branch pipe 4.

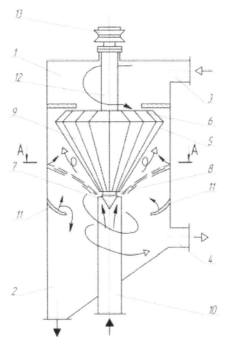

FIGURE 20.12 Dynamic gas washer [7].

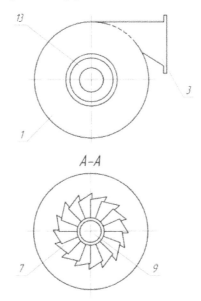

FIGURE 20.13 The top view and a cut{section} and.

20.11   THE WHIRLWIND APPARATUS WITH APPLICATION OF ULTRASONIC VIBRATIONS

The offered build will allow raising efficiency of clearing of aerodispersion mixes at the expense of superposition of ultrasonic oscillations in a zone of screw curling to 96 percent. The best results have been had at frequency of ultrasound 21.6 kc. Thus the water resistance was downgraded to 50mmwatercolumn. Experiments have shown that installation of the air swirlers executed with application of ultrasonic oscillations allows to expel cloging of channels by dispersoid adjournment.

The technical result secured by the invention, consists in an intensification of process of separation of aero dispersion gases.

The specified problem is solved because in the whirlwind apparatus containing the body, consisting of the screw twisting device and three chambers: the power separation, the cleared gas and dusty, exhaust tubes, and a discharge opening for sludge, unlike a prototype, in a zone of the screw twisting device magnetostrictive converters, thus the screw twisting device are installed is executed with dual-lead channels with the cavities of the right-angled cross-section decreasing in the direction of a stream in which zone two magnetostrictive converters, the ultrasonic oscillations connected to the oscillator are mounted.

The technical result secured by the whirlwind apparatus with application of ultrasonic oscillations, is expressed in raise of efficiency of clearing of gases and decrease in a water resistance of the device. Application of ultrasonic oscillations at clearing of gases allows to expel stagnation zones that does impossible an overgrowing of its internal elements by dust adjournment at work of the apparatus, it in turn promotes decrease in a water resistance of the device. Thus, the optimum operating mode of the whirlwind apparatus that gives the chance to expand a range of resistant to work at variable characteristics of a dust-laden gas stream is created and to raise productivity and efficiency of clearing of aerodispersion gases.

The essence of the invention is illustrated by the drawing:

On Figure20.14, the Cross-cut of the whirlwind apparatus with application of ultrasonic vibrations.

The whirlwind apparatus with application of ultrasonic vibrations according to Figure 20.15. Contains the body rolling the camera of power separation 1 in which the screw twisting device 2 with the diaphragm channel for removal of the cleared stream 3, connected in overhead on a course of a stream of a part with the chamber of the cleared gas 4 is installed. Over the chamber of power separation, the chamber 5, which is supplied by the point for supply of an entrance stream 6 is placed. From an outer side of the chamber of power separation, two magnetostrictive converters 7 supplied with points for input of ultrasound from the oscillator ultrasonic колебаний8 are had. For deductions of the cleared stream in the overhead and bottom part of the apparatus target points 9 and 10 are placed. Removal of the accumulated sludge is carried out through a discharge opening for deductions of sludge 11. In the bottom part of the apparatus, the observation port 12 and the sludge remover 13 is provided.

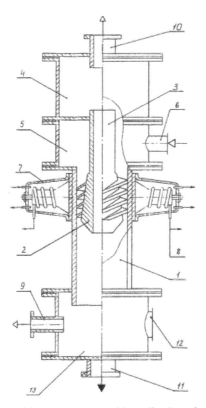

FIGURE 20.14 The whirlwind apparatus with application of ultrasonic vibrations [8].

The whirlwind apparatus works as follows.

The aerodispersion mix through the upstream end 6 arrives in channels of the screw twisting device 2 and under the influence of ultrasound from magnetostrictive converters 7 in the form of the twirled stream gets to the chamber of power separation 1. The dispersoid separated by a centrifugal force under the influence of ultrasound from magnetostrictive converters 7 gets to the sludge remover 13, coarsens and inferred through a discharge opening 11. The cleared gas through the diaphragm channel 3 gets to the chamber of the cleared gas 4 and is drained from the apparatus through a connecting pipe 10.

The offered build of the apparatus at the expense of use of ultrasonic vibrations allows to create favorable conditions for effective motion of a dispersion stream in the screw twisting device, expels stagnation zones that leads to raise of efficiency of separation and water resistance decrease.

Research of effect of design data of the chamber of power separation and the screw twisting device with ultrasound application on efficiency of separation.

Researches on studying of separation properties were spent on metal dual-lead vortex tube $D=125\cdot2.5$mm with a diameter on a thread of 175mm with a variable depth of thread from 10 to 25mmwatercolumn of cutting of 8.5, a conditional outlet angle of gas $\alpha =78°$. Superposition of ultrasonic vibrations for clearing of a gas mix was spent with the following fasted frequency: 18.5; 21.6; 22.1; 23.5 kc. The best results are had at frequency of ultrasound 21.6 kc.

The aerodispersion ftalo-air gas-vapour mixture was cleared of a dispersoid, at change of gas rates 196–250nm³/h and a dispersoid 0.62–23.0nm³ As a result of experiences have been installed that efficiency of separation in the whirlwind apparatus with the right-angled twisting channels makes 96 percent. Thus the air swirler water resistance was in limits of 50mm.water.column.

Experiments have shown that superposition of ultrasonic vibrations in the course of clearing of gas emissions allows to expel a driving of channels adjournment of a dispersoid and to raise efficiency of separation.

20.12   BUBBLING—SWIRLING APPARATUS WITH PARABOLIC SWIRLER

Proposed apparatus comprises cyclone, cylindrical chamber with inlet tube, tube to transfer slurry into slurry collector, axial sprinkler perforated all over its length and plugged at its outlet. Cylindrical chamber incorporates a pair of sequentially arranged swirlers. Swirler, first along gas flow, is rigidly fixed with axial sprinkler, while second swirler has centralbore and is jointed to cylindrical chamber walls tomake a gap for gas flow to pass to apparatus central zone. Said swirlers represent an elliptic paraboloid with its guide blades bent along helicoid to form curvilinear confuser channels.

Intensified gas cleaning due efficient centrifugal effects.

The bubbling—swirling apparatuswith a parabolic air swirler according to фиг.1 contains cylindrical camera 1c an entrance pipe 2, the axial sprinkler 3 perforated by holes of small diameter 4 and silenced with target end. In the cylindrical chamber 1, the pair of air swirlers 5 is consistently installed, and the air swirler on an entry of a gas stream is rigidly fastened to an axial sprinkler 3, and on an exit the air swirler is executed with the central hole, equal 0.2 diameters.

The cylindrical chamber, also it is connected to walls of the cylindrical chamber 1, forming a positive allowance for passage of a gas stream to the central zone of the apparatus. Directing guide vanes 6 parabolic air swirlers, 5 are bent on a helical surface and form curvilinear confuser channels 7. The cylindrical chamber 1 is affiliated with an inclination to low 8 c the help of flanges 9. Removal of dispersion particles is carried out by means of a pipe of an overflow of sludge 10 in the sludge remover 11.

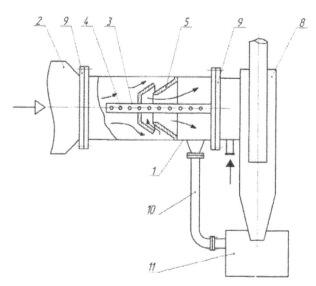

FIGURE 20.15   The bubbling—swirling apparatus [5].

The bubbling—swirling apparatus works as follows. Dusty gas is fed in the cylindrical chamber 1 on an entrance pipe 2. Simultaneously in an axial sprinkler 3, the irrigation water, which disperses on all volume of the cylindrical chamber from sprinkler holes arrives. The gas containing firm and gaseous impurity moves along walls of the cylindrical chamber 1 and is divided by means of an air swirler 5 into streams. Continuing a translational motion, these streams deviate a horizontal heading directionally parabolic a profile and get acceleration in curvilinear confuser channels without growth of turbulent and transverse pulsing.

After that, the gas stream bends around an air swirler on an entry, changing a traffic route, and initiates to be twirled in a positive allowance between air swirlers, forming turbulent gas-liquid a layer (mobile foam). Attaining a hole in an air swirler on an exit, gas passes through it and is inferred from the apparatus. The separated sludge is washed off by a liquid and by means of an inclination of the cylindrical chamber 1 is carried on an overflow pipe sludge 10 in the sludge remover 11.

The subsequent separation of slurry occurs in the cyclone separator 8, whence sludge also arrives in the sludge remover 11.

20.13   BUBBLING—SWIRLING APPARATUS WITH CONICAL SWIRLER

Invention can be used in wet cleaners of gas outbursts used in chemical, oil, and gas industries, etc. Proposed device comprises cyclone, cylindrical chamber with inlet tube, tube to allow sludge to flow intosludge collector, axial sprayer perforated over itsentire length and plugged at its outlet, and swirlers. Besides said cylindrical chamber incorporates plates arranged in vertical lines along gas flow. Swirlers are arranged staggered on plate surface and represent hollow truncated cones with their lather bases located in the plate plane. Surface of said cones is profiled along conical spiral. Plates are arranged to form zigzag-like gas passage.

Higher efficiency, efficient environmental protection.

The bubbling—swirling apparatus with a conic air swirler according to Figure 20.16 contains the cylindrical chamber 1 with an entrance pipe 2, the axial sprinkler 3 perforated by holes of small diameter 4 and silenced with target end.

The cylindrical chamber 1 is supplied by vertical plates 5, which

Are installed with formation of the zigzag channel 6 for passage of a gas stream. Air swirlers are executed in the form of hollow frustums of a cone 7, the big basis of plates lying in a plane 5. A surface of frustums of a cone profile on a conic spiral. The cylindrical chamber 1 is affiliated with an inclination to the cyclone separator 8 in terms of a flow of sludge by means of flanges 9. The Slope of the cylindrical chamber 1 is selected by practical consideration within 5–8, sufficient for sludge removal. Removal of dispersion particles is carried out by means of a pipe of an overflow of sludge 10 in the sludge remover 11. The liquid on an irrigation arrives on the point of feeding into of an irrigation water 12.

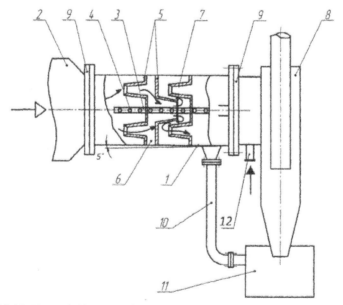

FIGURE 20.16   Bubbling—swirling apparatus with conical swirler [6].

The bubbling—swirling apparatus (Figure 20.16) works as follows. Dusty gas is fed in the cylindrical chamber 1 on an entrance pipe 2. Simultaneously in an axial sprinkler 3 the irrigation water on the point of feeding into of an irrigation water 12 which disperses on all volume of the cylindrical chamber from sprinkler holes arrives. The gas containing firm and gaseous impurity, moves along walls of the cylindrical chamber 1 and is divided by means of an air swirler into streams. Gas streams pass in positive allowances between plates 5, repeatedly changing the heading that leads to crushing of a liquid and an intensification of process of an interphase exchange. In positive allowances between plates, there is an intensive collision of corpuscles with irrigation water drops that promotes trapping of finely divided impurity.

The roll forming of an air swirler 7 on a conic spiral simultaneously with a stream twisting reduces possibility of a breakaway of dispersion particles and their ablation in a target pipe. Thus, drops of liquid with finely divided corpuscles with much smaller losses of ablation attain the sludge remover 11. The separated sludge is washed off by a liquid and by means of an inclination of the cylindrical chamber 1 is carried on a pipe of an overflow of sludge 10 in the sludge remover 11.

The subsequent separation of slurry occurs in the cyclone separator 8, whence sludge also arrives in the sludge remover 11.

20.14   INDUSTRIAL APPLICATION OF VORTICAL DEVICES

It is settlement—theoretical and design—design works were carried out with reference to operating conditions of manufacture hypochlorite calcium of Joint-Stock Company "Caustic"

On Joint-Stock Company "Caustic" the device is offered to be used for clearing smoke gases of furnaces of roasting 1.

According to the technological scheme, departing from the furnace of roasting 1 gases at temperature 550°C act in the bubbling-vortical device 2. Here on an irrigation 1–3 percent a solution of limy milk (pH = 11.5 − 12.5) move. Separated slime acts in a drum—slake 3; clarification and cooling of limy milk happens in the filter-sediment bowl 4: from which it feed on irrigation. The cleared gas stream smoke exhauster 7 is thrown out in an atmosphere.

Thus, introduction bubbling—the vortical device will allow to solve a problem of clearing of smoke gases with return of all caught dust in the form of slime in branches of slaking lime that provides captives manufactures.

Bubbling—the vortical device is supposed to mount in vent dust removal systems with the purpose of economy of the areas of industrial premises, thus hydraulic resistance of system does not exceed 500 Pa, power inputs on clearing of gas in 3 times below, than in known devices.

Application bubbling—the vortical device allows to achieve an intensification of process of gas purification with reduction of a gassed condition of air pool (Figure 20.17).

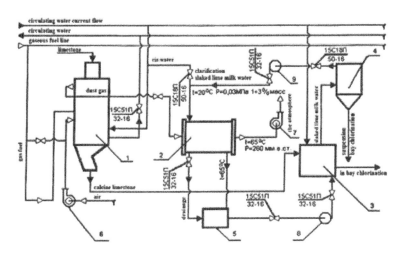

FIGURE 20.17   Flowchart of clearing of gas emissions.

20.15   CONCLUSIONS

Necessity of creation of high-efficiency mass transfer equipment has led to development of essentially new devices of parallel flow-vortical type. The basic directions of their application are:

Modernization of the existing columned equipment with a view of increase in productivity;   Creation new high-efficiency absorpritive and rectificative the columns working at raised {increased} and average pressure;Development of compact devices for operating{working} manufactures with the limited floor spaces and the manufactures removed from transport highways.

Now the basic variants of constructive registration of vortical devices are certain. Perfection of separate units with the purpose about the further intensification of process and expansion of a limit of working loadings proceeds. However questions of decrease{reduction} in metal consumption, maintenance of adaptability to manufacture of designs are solved not completely.

Intensive researches of hydrodynamics direct-flow and parallel flow-vortical, one-and two-phase streams as such streams are widely used and in other branches of techniques {technical equipment} are carried out. The saved up{saved} theoretical and experimental material allows verification to count the basic hydrodynamical parameters of vortical devices. But the hydrodynamics is insufficiently studied {investigated} at the raised {increased} and average pressure, i.e., in one of preferable areas of use of vortical devices.

Significant successes are reached {achieved} in the field of mathematical modeling process mass transfer in vortical devices. The received dependences enable theoretically to define {determine} influence of various factors on efficiency of process and to generalize experimental data. At the same time kinetics mass transfer at separate stages, for example at dispersion liquids, impact of drops, it is studied {investigated} insufficiently full.

Results of laboratory and bench researches show that vortical devices provide necessary efficiency of process and allow in increasing specific productivity of the columned equipment some times. Successful industrial tests of vortical devices once again confirm perceptivity of their use in a national economy.

KEYWORDS

- **Bubbling—swirling apparatus**
- **Efficiency**
- **Gas washer**
- **Rotoklon**
- **Swirler**
- **The patent**
- **Vortical devices**

REFERENCES

1. Usmanova, R. R.; Bubbling—the Vortical Device. The Patent of the Russian Federation the Invention No 2182843. May 27,**2002**. The Bulletin no 15.

2. Usmanova, R. R.; Bubbling—the Vortical Device with Adjustable Blades. The Patent of the Russian Federation the Invention No 2234358. August 20,**2004.** The Bulletin No 23.

3. Usmanova, R. R.; Bubbling—the Vortical Device with an Axial Sprinkler. The Patent of the Russian Federation the Invention No 2316383. February 10,**2008.** The Bulletin No 4.

4. Usmanova, R. R.; Rotoklon with Adjustable Sine Wave Blades. The Patent of the Russian Federation the Invention No 2317845. February 27,**2008.** The Bulletin No 6.

5. Usmanova, R. R.; Bubbling-Swirling Apparatus with Parabolic Swirler. The Patent of the Russian Federation the Invention No 2382680. February 27,**2010.** The Bulletin No 6.

6. Usmanova, R. R.; Bubbling-Swirling Apparatus with Conical Swirler. The Patent of the Russian Federation the Invention No 2403951. November 20,**2010.** The Bulletin No 32.

7. Usmanova, R. R.; Dynamic Gas Washer. The Patent of the Russian Federation the Invention No 2339435. November 20,**2008.** The Bulletin No 33.

8. Usmanova, R. R.; The Whirlwind Apparatus with Application of Ultrasonic Vibrations. The Patent of the Russian Federation the Invention No 2482923. May 27,**2013** The Bulletin No 15.

INDEX